Fossils in the Field

Information Potential and Analysis

ROLAND GOLDRING

▶■▶ Longman
■■■ Scientific &
■■■ Technical

Copublished in the United States with
John Wiley & Sons, Inc., New York

Longman Scientific & Technical,
Longman Group UK Limited,
Longman House, Burnt Mill, Harlow,
Essex CM20 2JE, England
and Associated Companies throughout the world.

Copublished in the United States with John Wiley & Sons, Inc., 605 Third Avenue,
New York, NY 10158

© Longman Group UK Limited 1991

First published 1991

British Library Cataloguing in Publication Data
Goldring, Roland
    Fossils in the field: information potential and analysis.
    1. Stratigraphy. Correlation analysis. Quantitative methods
    I. Title
    560
    ISBN 0–582–06261–6

Library of Congress Cataloging-in-Publication Data
Goldring, Roland, 1928–
    Fossils in the field: information potential and analysis/Roland Goldring.
        p. cm.
    "Copublished in the United States with John Wiley & Sons, Inc., New York."
    ISBN 0–470–21679–4
    1. Fossils.   2. Palaeontology—Field work.   I. Title.
QE711.2.G65   1991
560—dc20   90–33994
                                                                                         CIP

Po 2959

Set in 10/13 Plantin

Produced by Longman Singapore Publishers (Pte) Ltd.
Printed in Singapore

Contents

Chapter 4   Taphonomy

Preface

Fossils comprise a unique data set, providing answers to a range of questions, and having application to numerous theories. As such, data collection is pertinent to disciplines as diverse as oceanography, geochemistry and geophysics, as well as to palaeontology. The aims of this book are therefore to provide a basis for evaluating the information potential of fossiliferous sediments, and then to give an outline of the strategy and tactics which should be adopted whilst gathering the information in the field. Some readers will be disappointed that the text does not provide a simple user-friendly list of jobs to be done at a site. This is because of the numerous information categories held in fossiliferous rocks; and because, even within a single facies, such as shallow marine sandstones, the recoverable information is so variable (depending on the tectonic style, diagenetic history and type of weathering).

The text assumes that the reader will have a general knowledge of the plant and animal phyla, and that it will be used in conjunction with a book on fossil recognition. No keys for identification are included (*but see* Appendix B) because they are not feasible at this level for vertebrate or plant fossils, and trials with undergraduate classes have shown that if students had enough knowledge to use the keys then they did not need them. Furthermore, keys cannot cope with many of the modes of preservation, e.g. a discrete spar-filled cephalopod camera. The reader will probably have carried out some mapping in the field area, or made a site investigation, and will be familiar with the geological time scale and basics of sedimentology. The book is not designed as a simple field handbook but as a guide, prompt and information source to be used at base, at the exposure and in the laboratory.

The book collates many facts, normally only available in reference collections and libraries, without recourse to which it may not be possible to pursue a particular line of investigation in the field. For example, approaches in field observation, largely influenced by geological theory, also draw on developments in other branches of science. In palaeontology, it is the theoretical advances in biology and ecology which have made the most impact; also significant, are advances in sedimentology, tectonics, geochemistry and biochemistry. The text thus aims to make the reader aware of relevant concepts from these rapidly developing disciplines. Furthermore, important as fieldwork is, there are many geological happenings, such as an anoxic or iridium event, affecting the microbiota, that leave no clear visual evidence at the exposure, and such information must be collated from elsewhere. One may be suspicious, one may speculate, and the answer may be forthcoming from later laboratory investigation of samples that are precisely located. Thus a good many facts are included in the text, but due to limitations of space, full documentation is not permissible. However, where possible topics, such as trace fossils, which are interdisciplinary in nature, are cross-referenced between several chapters.

Although this book is essentially concerned with fossils in the field, some of the most exciting and important finds in recent years have been made in our heritage collections: national museums, university and school collections, Survey archives and private collections. In the past, as now, collectors may have been unsure of the identity of a particular specimen, or were puzzled by an odd association or type of preservation; or thought it odd that such and such a species should have been found where it was. It was not always prepared as expertly as it might be nowadays, but it was labelled and deposited. What a treasure chest these collections are, especially for palaeontologists, palaeobotanists and sedimentologists.

Acknowledgements

This text has had a long gestation, but it was Rose Dixon (formerly of the Open University Press) who got the project under way. To Ian Francis and Lesley Evans, of Longman, I am immensely grateful for their demanding but always constructive criticism and real interest, especially in the reconstruction of the text. Action by the Geological Society of London has much delayed publication but many Fellows and others have generously gone over, at various stages, smaller or larger sections or made useful suggestions. In particular, I thank Richard Bromley, Howard Brunton, Euan Clarkson, Michel Gruszczynski, Jake Hancock, John Hudson, Jim Kennedy, Sue Kidwell, Clive McCann, Tim Palmer, John Pollard, John Rippon, Andrew Scott, Brian Selwood, Peter Skelton, Bob Spicer, George Warner, Barry Webby and Mark Wilson. Special thanks go to past and present students, Tom Aigner, Omer Ali, Dan Bosence, Paul Bridges, Mike Brookfield, Peter Crane, Brian Daley, Holger Hogrefe, Quentin Huggett, Andy Hughes, Ed Jarzembowski, Thomas Junghanss, Ray Milbourne, Guy Plint, John Simpson, Dave Stewart and Maurice Tucker, for their forthright comments and for contributing in ways that only students can. Thanks also to the classes of '87 and '90 for acting as guinea-pigs. I am also glad to acknowledge the photographic work of Jim Watkins and much of the early typing by Elizabeth Wyeth.

Dan Bosence kindly supplied the original for Fig. 2.13 and Karl Flessa that for Fig. 2.15. Figures 3.4, 6.24 are based on or include unpublished figures by Robert Riding and John Pollard. The following gave material used for figures: Bill Baxendale (4.10), L I Salop (4.13), Gunter Freyer (4.23), Andy Hughes (6.5), John Evans (6.6), Alan Kendall (6.12), David Stephenson (6.14), Mike Mayall (6.25), George Best (7.11), Stephen Eager (9.3).

We are indebted to the following for permission to reproduce copyright material:

The author, Dr. D.J. Bottjer for fig. 7.8 (Droser & Bottjer, 1990); International Association of Sedimentologists for fig. 3.5 (Bridges & Chapman, 1988); Schweizerbart'sche Verlagsbuchhandlung for fig. 7.7 (Shäfer, 1969); Société Géologique de France and the author, Professor C. Pomerol for fig. 8.2 (Pomerol & Cavelier, 1986); Springer Verlag and the author, Dr. E. Futterer for tables 7.3 & 7.4 (Futterer, 1982).

Whilst every effort has been made to trace the owners of copyright material, in a few cases this has proved impossible and we take this opportunity to offer our apologies to any copyright holders whose rights we may have unwittingly infringed.

Safety in the field

'Forewarned is forearmed'

1. You are responsible for your own safety. Ascertain what essential equipment is required, and what are the safety procedures for the area to be visited. Follow points 2 and 3 for northwest Europe.

2. *Essential equipment:*
    Map and compass; watch; whistle; small torch; First Aid kit; rain and wind-proof gear, including a brightly coloured item; warm clothing reserve, including polythene survival blanket; sun hat; high-calorie food reserve; water container; walking boots; hard hat and goggles; knife.

    General procedure:
    Listen to the daily weather forecast (including wind direction), which may determine where it is prudent to work.

    Take account of the time and height of tides when planning coastal work.

    Write down each day your approximate route, working area and time of return, and leave it for others to see.

    In worsening conditions do not hesitate to turn back if it is still safe to do so.

    If you get lost, disabled, benighted, or cut off by the tide, do not hesitate to stay where you are until conditions improve, or you are found. Supposedly short cuts under such circumstances can be lethal.

3. *Distress code:*
    On mountains – 6 long blasts/flashes/shouts/waves in succession repeated at minute intervals

At sea signal SOS – 3 short/3 long/3 short blasts or flashes, pause and then repeat.

NB Rescuers reply with 3 blasts/flashes repeated at minute intervals.

References to safety and conduct in the field

A Code for Geological Field Work Geologists' Association, London

A Code for Geological Field Work Geological Survey New South Wales Mining Museum

Code of Practice for Geological Visits to Quarries, Mines and Caves Institution of Geologists, London

Code of Practice for Scientific Diving 1979 Underwater Association

Mountain Safety: Basic Precautions Climber and Rambler, London

NICHOLS D (Ed) 1983 *Safely in Biological Fieldwork – Guidance Notes For Codes of Practice* Institute of Biology, London

Safety on Mountains 1975 British Mountaineering Council, Manchester University

THOMASON J G (Ed) 1983 *Guidance Note-Safety in Fieldwork* Natural Environment Research Council, London

1

Principles and classification

'The best geologist is the one who has seen the most rocks'
H H Read, J V Watson (Imperial College, London) in Preface to
Beginning Geology (Macmillan), slightly emended

All geologists need to have an appreciation of fossils. The main object of this text is therefore to draw attention to the usefulness of fossils, as seen in the field, to individuals who will be looking at fossils with different objectives, and with different backgrounds and experience. This chapter discusses first the various categories of information that can be obtained from fossils, and then outlines a classification of fossiliferous sediments that reflects these information categories. The object is, at an early stage of any investigation, to assess the value of a site, and how the information potential may respond to vertical and lateral change.

A group of palaeontologists at a fossiliferous site pursue a number of different goals. Some are concerned with determining the biostratigraphy, some with the palaeoecology, others are searching for especially well-preserved material. Many follow carefully prepared sampling programmes, in order to collect fresh material for micropalaeontological or geochemical analysis. Similarly, the sedimentologist, having logged a section and determined the lithologies and current directions, has to consider the available biological data to make a convincing palaeoenvironmental analysis. It is to the palaeontologist, palaeoecologist and sedimentologist that this text is principally directed: how to recognize the potential of fossiliferous sediments and analyse the sediments to best advantage. The task of palaeoecologists and sedimentologists concerned with palaeoenvironmental interpretation is particularly demanding because they have to be at the same time palaeontologists, palaeoecologists, hydraulic sedimentologists, sedimentary geochemists and sedimentary mineralogists. The best way to deal with a particular site is to take each of these roles in turn; but, nevertheless, keep in mind the way in which each integrates with the other.

The biggest hurdles to be faced in the organic world are the colossal diversity of life, today and in the past, and the unpredictability of evolution. Added to this is the problem of determining to which group of animal or plant a fossil or fossil fragment belongs, especially when only, for instance, an impression is available, or it has been deformed tectonically.

If the immediate urge on locating a fossil is to extract it, then pause and consider what information might be forthcoming from a careful examination of its position and attitude in the sediment. A detective investigation is under way and there is seldom a simple answer and rarely can every line of information potential be completely followed through.

1.1   Categories of information

There are eight main categories of information that can be obtained from fossiliferous sites:

A	SYSTEMATIC & MORPHOLOGICAL	Information about the morphology and organization of hard and soft parts that can be used in identification and classification.
B	PHYSIOLOGICAL	Information that can lead to an understanding of the original function (physiology) of the fossils. Good information is likely to be provided by material fossilized more or less in the position in which it lived (autochthonous), and where soft tissue has been preserved.
C	EVOLUTIONARY	Information that can lead to an appreciation of the evolutionary position and evolutionary pattern of the fossils within a particular unit. The pertinent question is whether, in a sequence, it will be rewarding to sample sequentially to gain an appreciation of evolutionary changes. Stratigraphic breaks and facies changes need to be minimal.
D	ENERGETIC	Information that can be used in assessing ancient productivity and ecological energetics. Preservation of growth laminae is important for any work in this field. (*Energetics*: energy transformations within communities.)
E	ECOLOGICAL	Information that can be used to determine ancient ecological relationships and palaeoenvironments. Autochthonous material will be particularly useful,

together with indications of trophic niche (such as predation), inter-specific relationships, relationships to substrate surface etc.

F HYDRAULIC, STRATINOMIC & YOUNGING	Information that can be used for determination of the hydraulic regime, stratinomic aspects and as way-up, younging (geopetal) criteria. This information differs from preceding categories in that autochthony is not necessarily involved, and it is generally the hydraulic and stratinomic attributes of the fossils that are of interest. (*Stratinomy*: the processes between death and final burial.)
G STRATIGRAPHIC	Information that can be used for stratigraphic correlation. Here degree of autochthony or allochthony, fragmentation and dissociation is less important providing there has not been significant reworking, and providing that identification is still possible. A fragment of a biostratig-raphically important taxon that still retains its hallmark will be sufficient.
H DIAGENETIC	Information that can be used to appreciate diagenesis and diagenetic history. It is shape, structure and composition of fossils that are the intrinsic variables that relate to the chemical and compactional changes in sediments, which largely follow completion of stratinomic processes and precede later diagenetic change associated with tectonic history. (*Diagenesis*: the changes, chemical, physical and biological, that occur after initial deposition. *See* Appendix F.)

1.2   Principles

Two principles, in particular, are involved when investigating fossiliferous sediments:

1. Different groups of fossils have different information potential, either because of their inherent attributes, or because of taphonomic considerations.

2. Different types of fossiliferous sediment yield different categories of geological information.

The first principle is one that is generally appreciated. Systematic and stratigraphic classifications of fossils are used at an early stage of geological experience. Taking some examples: the relatively small amount of morphological information that can be obtained from a fossil gastropod shell contrasts with the much greater amount that is generally available from an echinoid test. Poor preservation may further reduce the information from a strongly recrystallized gastropod shell. Or, compare the high stratigraphical value of many graptolites with the low value of contemporary bivalved molluscs. Similarly, it is shells with moderate convexity and bilateral symmetry that are the most useful types in the determination of flow characteristics. Also, benthonic calcareous algae are more useful in environmental interpretation than are many other groups.

The second principle may be illustrated by scanning Fig. 1.1 or, more closely, a section in Upper Jurassic (Oxfordian) sands and limestones (Fig. 1.2; *see also* §2.1). In the basal sands, vertical and horizontal arthropod burrows (*Ophiomorpha nodosa*) are useful environmental indicators; while more precise stratigraphical information may be obtained from thin muddy partings, yielding spores, pollen and occasional microplankton. The palynofacies supports interpretation of the sands and muds as representing an inshore depositional environment. The only moderately good stratigraphical information from the upper, calcareous part of the section is from relatively uncommon and poorly preserved ammonites from the oolites. No ammonites have yet been discovered from the coralliferous part of the section and its age is somewhat uncertain. Bivalves at the coarse-grained, shelly base to the transgressive carbonates often show preferred orientation and provide information on the nature of local erosional environments. The coralliferous unit, in which the branching coral *Thecosmilia* dominates, has an abundant and well preserved associated biota, including regular echinoids and bivalves. Examination of the wackestone matrix reveals small gastropods, brittle-star 'vertebrae' and small thecideidine brachiopods (Fig. 6.3). The branching and massive corals have provided opportunity for growth and productivity studies. The muddy oolite also displays an autochthonous/parautochthonous biota of infaunal bivalves and the irregular echinoid *Nucleolites scutatus*. In this sequence the information potential changes rapidly corresponding with the facies changes.

1.3   A classification of fossiliferous sediments

Schäfer (1972) proposed a marine sediment classification, applicable to ancient fossiliferous environments. Evolving from a similar ecological work by Hermann Schmidt, the classification showed that stratification and completeness of the sedimentary record (for benthic organisms and sediments alike)

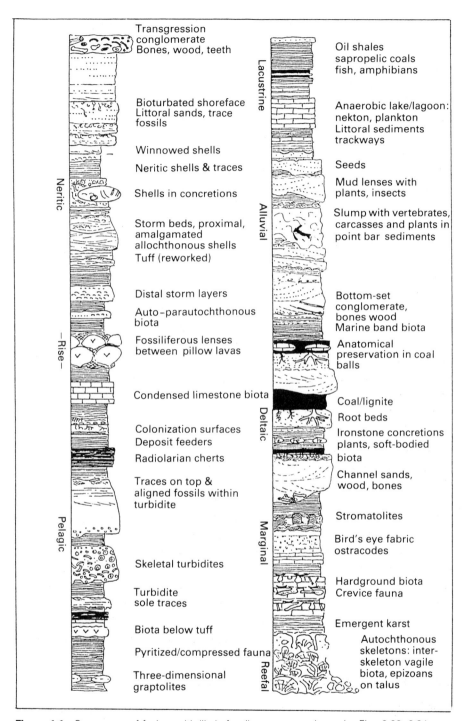

Figure 1.1 Sequences of facies, with likely fossil occurrences (*see also* Figs 6.23, 6.24, 6.28)

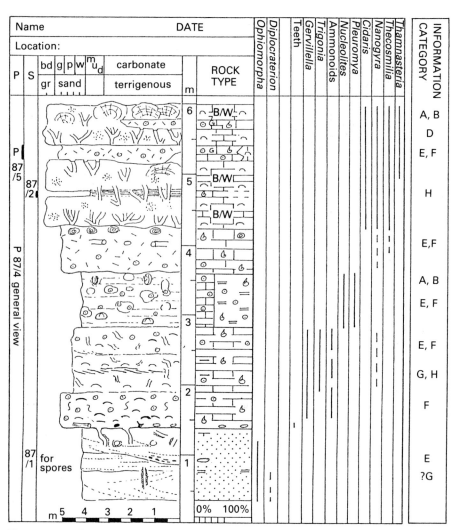

Figure 1.2 *Left page.* Views of quarry face at Stanford Quarry, Faringdon, Upper Jurassic and graphic log of section. Column on right is indication of categories of information obtainable from the various beds. Abbreviations: P – scene photographed, S – sample, with number. Sediment classification: bd – boundstone, g – grainstone.

p – packstone, w – wackestone, gr – gravel. Dots at 3.6 m.
Right page. Upper left, from print of thin-section of *Fungiastrea arachnoides* showing diagenetically-enhanced zones of thick septa, and light zones with poor preservation. Upper centre, growth rates of three corals in the Corallian with mean and standard deviation, determined from measurements of diagenetically enhanced zones (after Ali, 1984). Upper right, generalized representation of section to illustrate coral growth rate and minimum time of formation. Below. Reconstruction of two palaeocommunities at Stanford Quarry (adapted from Fürsich, 1977, who used the term association): (1) *Nucleolites scutatus* community with its (skeletal) trophic nucleus (the species that account for 80% of the fossils). This is how Fürsich would reconstruct the nodular, muddy oolite 3 m above base; (2) *Gervillella aviculoides* community, corresponding to the lower carbonate beds, with trophic nucleus and orientations of *Gervillella* measured on fallen blocks, with number of specimens. At foot of diagram rarefaction (or sampling curve) for the two communities:

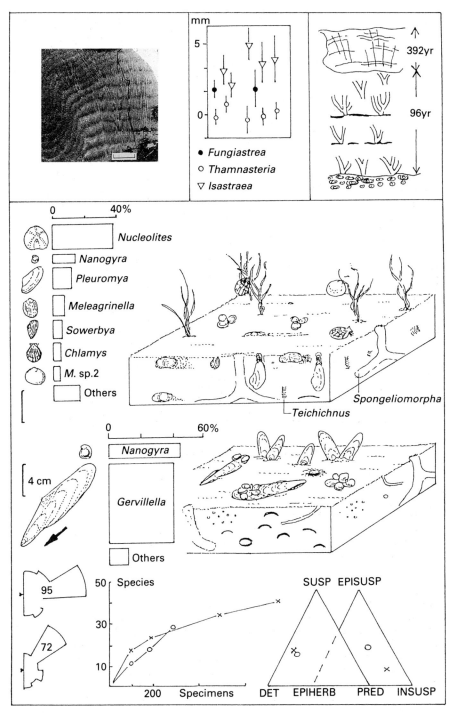

x – *Nucleolites*, o – *Gervillella* showing plot of number of species against number of specimens, and triangular diagrams for the two communities. Left, feeding habits and right, substrate niches.
SUSP: suspension; EPISUSP: epifaunal suspension; DET: detritus; EPIHERB: epifaunal herbivore; PRED: predator; INSUSP: infaunal suspension (for details, see §6.2.5)

were closely related to the degree of oxygenation. Schäfer's scheme (slightly modified in Fig. 6.11) is extended to 13 broad types of fossiliferous sediment to identify more closely the categories of information potential. Of primary concern is, whether or not benthic organisms and traces are present; since their presence, if more or less in life position, indicates at least a degree of oxygenation at the substrate. Dominance of nektonic and/or planktonic fossils will generally suggest accumulation under anoxic or dysaerobic conditions, though this may not be readily resolved. In the fossil record, cephalopods, graptolites and pelagic trilobites are some of the most useful stratigraphic markers. Chalk is unusual in that, though the substrate was for the most part richly colonized, as evidenced by the high degree of bioturbation and skeletal remains, the sediment 'rain' of coccoliths and planktonic foraminifera is of stratigraphic importance. Fossils of subaerial plants are mostly concentrated in facies quite distinct from animal fossils. The primary, if empirical, criterion of the classification is whether the site, or sedimentary unit, yields dominantly benthonic, dominantly nektonic and/or planktonic, or dominantly plant fossils.

Brett and Baird (1986) have coined the useful term taphofacies (§4.1) for sedimentary rocks characterized by particular, sometimes unique, combinations of preservational features; the result of similar taphonomic pathways. However, in their discussion of trilobite taphofacies Brett and Baird were naturally most concerned with stratinomic pathways, the sorting, reworking and damage that may occur before final burial, rather than the early diagenetic, mainly chemical, aspects. In practice, stratinomic and diagenetic aspects are analysed separately at outcrop and the stratinomic aspects cannot, of course, be separated from hydraulic considerations. How stratinomy and early diagenesis reflect intrinsic factors such as skeletal mineralogy, dissociation, disarticulation and fragmentation, as well as extrinsic factors, such as sedimentation rate and burial processes, are discussed more fully in Chapter 4.

The second criterion in the classification is the influence of taphonomic processes on the facies. Later diagenesis and tectonic history are too variable to be predictable from the depositional environment though some sedimentary facies are, for instance, more prone to de-dolomitization, or particular styles of tectonic deformation, than others. Information potential is also influenced by the degree of stratigraphic condensation or expansion. Hardgrounds are by definition condensed and pelagic shales and pelagic nodular limestones are also typically so; slumps tend to be expanded and associated with steep depositional gradients. But the organization of shelly fossils in turbidites is generally similar to their organization in shallow water storm-generated sediments, such that the information potential is alike in most respects.

1. Benthonic fossils predominate in types 1–6.
2. There are two main types of fossiliferous deposit where pelagic or epiplanktonic fossils are the principal fossils (7–8).

3. Five types of fossiliferous deposit where subaerial plants are the principal fossils (9–13).

This classification cannot be exhaustive; there are intermediates and there are other types of skeletal accumulations associated with biogenic stratification and biogenic graded bedding described from modern environments (§3.3). While it is useful to distinguish subaerial (mainly vascular) plants in volcanic associations, it is less important to separate skeletal accumulations in such facies for their information potential. Furthermore:

1. Each type of deposit may be dominant in a sedimentary sequence, or may be present as only a minor facies. For instance, bituminous shales (7) may be a minor facies in a sequence dominated by rather unfossiliferous turbiditic muds and silts.
2. Any of the types distinguished may contain fossils of such palaeontological significance, either because of the concentration of fossils or because of unusual tissue preservation, that a fossil-ore (Fossil-Lagerstätten) is represented (§6.3).
3. Many of the fossils that are most typical of one, or other, type of fossiliferous sediment may be found in another. For example, anatomically preserved plants occasionally occur in marine shales.

Most of the types of fossiliferous deposit described below are referred to in the examples given in Chapter 2.

1.4   The next stage

Following recognition of the different categories of fossiliferous sediment, consideration should be given to the investigation in progress. Both the time available and the amount of detail required will place limitations upon the course of study. Depending upon what information is of most concern, each category will require a rather different approach. For example:

- If the only significant results are likely to be of a hydraulic nature, then emphasis should be placed upon the sedimentological aspects of the site (see Chapter 7).
- Where contributions to knowledge of the morphology and functional morphology of a certain taxa may be determined, it may be more prudent to concentrate the field study on particular beds or intervals of the section under examination. Attention should also be given to the sedimentological setting and associated biota.
- In instances where environmental interpretation is the primary objective, the ecology of each taxa (autecology) should be determined initially, with further thought being given to hydraulic aspects.

A sophisticated study of palaeoenergetics would require a carefully prepared programme, and is beyond the scope of this book.

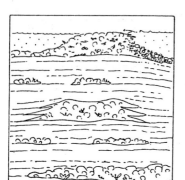 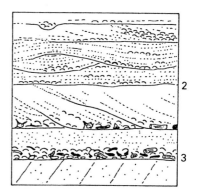

Table 1.1

1 Autochthonous buildups, ranging from stromatolites to small bryozoan mounds to oyster banks to barrier reefs, where biogenic processes are dominant. (Example § 2.1)	2 Well-stratified sediments (siliciclastic or calcareous), rapidly deposited, with hydraulic processes and products dominant. Ranging from fluvial sands, bioclastic shelf limestones and sandstones, storm beds to bioclastic turbidites and fossiliferous tuffs. (Examples § 2.1, 2.2)

A *Systematic & Morphological*

B *Physiological*
Good to excellent for many structures, but often loss of aragonite and/or dolomitization. Potential information on astogeny and life attitudes, though fossils may be difficult to release from matrix

Good to excellent, but morphological information may have to be pieced together from dissociated and broken material. Preservation can be good, though is often mouldic

C *Evolutionary*
Generally poor

With high abundance of fossils, potentially good, but size sorting typical, or sediment washed

D *Energetic*
Potentially good e.g. corals, oysters, depending on amount of diagenetic alteration

Poor or difficult to resolve, because of sorting and mixing (time-averaging)

E *Ecological*
Potentially good, especially for inter-relationships of biota

Mainly relates to environment(s) of provenance

F *Hydraulic & stratinomic*
Growth form may be indicative. Also talus within buildup and around margins may provide important information

Generally informative from sedimentary structures and skeletal fabrics

G *Stratigraphic*
Can be poor. Age relationships often best established from underlying and overlying strata

Variable and often difficult to resolve because of reworking and introduction of derived material

H *Diagenetic*
Complex early diagenesis, ± meteoric, vadose or phreatic cements, dissolution, replacement. Dolomitization common

In siliceous sequences where sedimentation rapid, ± early carbonate dissolution leading to moulds/replacements. Compaction low

In calcareous sequences early cementation often leads to good preservation. In regressive sequences meteoric cements likely

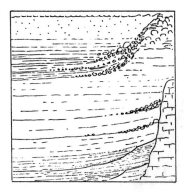

Table 1.1 (cont.)

3 Lag concentrates at unconformities and discontinuities, with hydraulic processes dominant. Can also include fissure fills. (Example § 2.4)

4 Mass flow, debris flow and talus deposits generally representing transport from shallow to deep water, from reefs or associated with submarine fault scarps. The biota was rapidly buried and records elements often absent or poorly preserved in the original environment. Examples include the Burgess Shale (Middle Cambrian) and Cowhead Breccia (Ordovician)

A–D As for (2)

A–B Good to excellent, and may include preservation of soft tissues

C Often very important source of information

D–E Difficult to determine because of redistribution

E More difficult to resolve than in (2) because of mixing, sorting and damage

F Generally distinctive

F Clear from sedimentary structures

G, H As for (2)

G Generally good but mixing can pose problems. Clasts may be older or younger than matrix

H Diagenesis likely to be complex, possibly involving predisplacement aspects

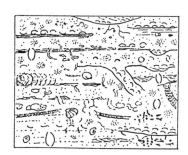

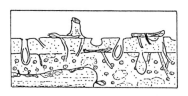

Table 1.1 (cont.)

5   Internally poorly organized (generally bioturbated, often heterolithic) sandstones and sandy mudstones, mudstones, limestones (pack- to wacke-stones), also some micritic limestones (most coccolithic chalks). Typically marine shelf sediments (also lagoonal and lacustrine) formed under mixed biogenic and hydraulic processes. Abundance of skeletal fossils influenced by net sedimentation rate so that 'end' members are shell and bone gravels (starved of sediment) or poorly fossiliferous mudstones (starved of fossils). (Example § 2.4)

6   Hardgrounds of varous types and depicted here by planed Jurassic shallow water hardground, and irregular Cretaceous Chalk omission hardground. (Example § 2.3)

A   *Systematic & morphological*

B   *Physiological*
Good to excellent for body and trace fossils, but disturbance or breakage by bioturbation likely

Can be good for encrusting and boring biota, especially in crevices and within burrows/borings (coelobites). Within hardground preservation good (e.g. preservation of aragonitic biota) when cement early

C   *Evolutionary*
Potentially good, but may be difficult to obtain large samples of a taxon

Generally not applicable

D   *Energetic*
Potentially good, but requires careful analysis

Difficult to interpret because of high loss of information

E   *Ecological*
Potentially good, with autochthony/parautochthony common

Specialized but good, especially for succession of borers and encrusters

F   *Hydraulic & stratinomic*
Can be difficult to establish, especially if primary structures destroyed by bioturbation, though way-up generally readily determined

Variable, but succession of colonizers may provide clues e.g. truncated borings

G   *Stratigraphic*
Generally good to excellent from macro- and microbiota

Care required because of incomplete sedimentary record

H   *Diagenetic*
Early phosphatic, sideritic, pyritic concretions typical. Early cements tend to be patchy. Compaction high in muddy sediments, ± preservation of aragonite

Compaction minimal, very variable cements especially in planar hardgrounds, where meteoric processes likely

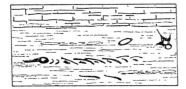 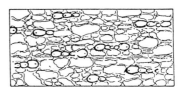

Table 1.1 (cont.)

7 Well laminated and often bituminous mudrocks, thin bedded limestones (Plattenkalk), also radiolarian cherts and diatomites. (Example § 2.6)	8 Nodular micritic limestones, the most typical being those packed with cephalopods (Ammonitico Rosso, Cephalopodenkalk)
Often exceptionally good to extraordinary	Often difficult to determine because of early dissolution of aragonite and calcite
Potentially excellent, but compression may hinder evaluation	As above
Not applicable because generally little or no benthos	Not applicable
Generally good	Needs careful analysis because of possible reworking and condensation
Seldom applicable, but if preferred orientation or imbrication present, then substrate not continually anaerobic. Way-up evidence often absent. Minimal stratinomic disturbance	As above
Generally very good from micro- and macro-biota	Generally good, allowing for preservational problems
± high compaction, pyritization (disseminated to nodular), skeletal dissolution common, ± preservation of organic matter (including soft tissue)	Variable early diagenesis depending on water depth, i.e. ± within photic and aerobic zones. Early cements (marine) common, ± skeletal dissolution

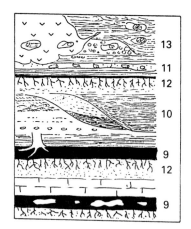

Table 1.1 (cont.)

---

9 Peats, lignites and coals. Coal balls and other indurated peats are particularly important. These are the plant equivalent to type 1

10 Well-stratified (often laminated) fluvial or lacustrine mudstones e.g. Mazon Creek, siltstones and limestones (e.g. Green River) abandoned channel fills and point-bar sediments. Early diagenetic concretions important. The plant equivalent of type 2 because of abundance of material

A *Systematic & morphological*

B *Physiological*
Generally poor (except for pollen, spores and in concretions). Arthropod cuticle may be common, though often fragmentary

Generally good for leaves as compressions of impressions

C *Evolutionary*
Good for microbiota

D *Energetic*
Potentially good if rate of accumulation can be determined

Difficult to determine

E *Ecological*
Distinctive but small-scale variation likely

Useful, but mixing may have occurred

F *Hydraulic*
Generally not applicable

Important to determine for E

G *Stratigraphic*
Generally poor (except from pollen, spores, and in concretions)

Generally useful from micro- and micro-flora.

11 Terrestrial slump deposits, often massive mudstones

A – G
As 10 but material generally less compressed.

Table 1.1 (cont.)

| 12 Root beds and autochthonous stumps, fossil forests | 13 Volcanigenic associations: lavas, tuffs and volcanically derived sediments, with calcification in basaltic terranes and silicification in rhyolitic terraces e.g. Midland Valley, (Scotland), Florissant Formation (Colorado). (Several preservation modes likely; concretions most important. Note also volcanic associated mud-flows of type 11.) |

Root beds	Fossil forests	
A–D Poor	A–D Variable, depending on mode of preservation	A–G Can be optimal after careful site analysis but ecological relationships may be strongly biased

E Good

F Not applicable

G Spores and pollen often oxidized in fossil soil

References

ALI O E 1984 Sclerochronology and carbonate production in some upper Jurassic reef corals. *Palaeontology* **27**: 537–48

BRETT C E, BAIRD G C 1986 Comparative taphonomy: a key to paleoenvironmental interpretation based on fossil preservation. *Palaios* **1**: 207–27

FÜRSICH F T 1977 Corallian (Upper Jurassic) marine benthic associations from England and Normandy. *Palaeontology* **20**: 337–85

SCHÄFER W 1972 *Ecology and palaeoecology of marine environments* University of Chicago Press, Chicago, USA, 568 pp

2

Examples and opportunities

'There is still much to be done, times are exciting, and new ideas, tools, techniques and theories ready for new and old sites'

This chapter presents a few thumb-nail descriptions of how fossiliferous sites have been, and may be, investigated, and how modern environments can be used to study taphonomic, ecological and sedimentological processes and products. The ancient sites have been chosen for their information potential and typicality. Some have had scarcely any modern work done on them; others have been the subject of detailed investigation. Numbers refer to the classification of fossiliferous sites given in Chapter 1.

2.1 Mesozoic intertidal sands, neritic limestones and a coral bank (types 1, 2, 3, 5, 6)

Stanford Quarry (Figs 1.2, 2.1), a few kilometres to the west of Oxford, England, is an extensive shallow quarry, formerly worked for the fine-grained sand which is of good mortar quality and a binder for bitumen (lately used for the new runway in the Falkland Islands, South Atlantic). The strata, of Upper Jurassic age, dip gently southwards. Here, the objectives of the investigation were to gain an appreciation of the depositional environment of the sands in order to predict their distribution laterally for extending the workings, and to carry out a sedimentological and palaeoecological investigation of the limestones and coralliferous unit; the latter being the only good exposure of such a facies in southern England. A reconnaissance survey, lasting at least an hour, showed that there was extensive lateral change along the conserved wall, revealing stratigraphically variable (Fig. 1.2) information potential. (As part of this initial investigation estimates were made of the materials needed for making peels from the loose sand.)

The main site investigation was tackled in two stages. Firstly, the sands were mapped in detail at scales of 1:10 or 1:20, and the facies dissected with a trowel,

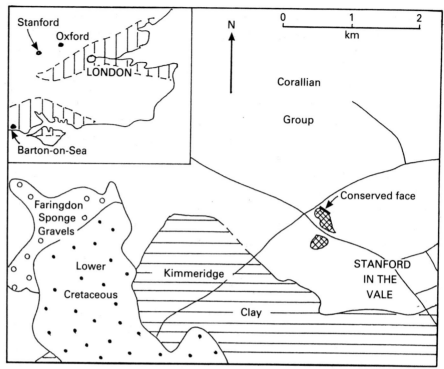

Figure 2.1   Geological sketch map and location of the conserved face at Stanford Quarry, near Stanford in the Vale, England. Inset also shows location of Barton-on-Sea (§2.4) and Tertiary basins of southern England. Note, lower left, unconformable Faringdon Sponge Gravels and Lower Cretaceous (§6.3)

and directions of foresets measured and recorded. Trace fossils were drawn to scale and further investigated by dissecting. Often, more detail could be observed by searching for wind-etched surfaces. Sections were photographed in colour and in black and white. Fresh samples (100 g) of the muddy laminae were collected from identified horizons for palynofacies analysis. Acetate peels (each taking several hours to prepare) were made of selected areas, and samples taken for grain size analysis and clay mineralogy.

In the second stage of the investigation, the limestones were mapped and logged, paying particular attention to the nature of bed contacts. Traced laterally, the slightly irregular erosion beneath the coralliferous unit could be detected. The lenticularity of the shelly limestones (bio–oosparites) within the coralliferous unit was also investigated.

A programme of alternating field and laboratory investigation concluded the study. This allowed some elements of the fauna to be identified in the field from British Museum handbooks, but most, in the laboratory from monographs. At this time oriented samples (minimum dimension 10 cm) were collected for slabbing and thin-section or acetate peel analysis, the position of each sample being recorded on the logs.

Examination of the coralliferous unit took considerable time. The phaceloid corals are mostly preserved as moulds in a cemented wackestone matrix. Their morphology was sketched, and blocks of the phaceloid and massive corals collected for examination and sectioning. Fresh samples of the clayey seams were collected for X-ray fluorescent (XRF) analysis and bags of the weathered clays collected with a knife and fine trowel for wet sieving. These were found to have a rich microbiota including thecideidine brachiopods (Fig. 6.3), small gastropods and brittle-star elements which would be included in the reconstruction of the unit. Later, specimens in regional and national collections were examined, and discussions led to closer identification and better knowledge of the existing literature.

The coralliferous section represents a bank rather than a reef, but a similar approach can be used for any largely autochthonous accumulation. The bedded limestones can also be matched throughout the Phanerozoic.

2.2   Upper Devonian shelf sandstones and shales (type 2)

The objective of this study was to assess the macrofossils of the Pilton Formation, some 700 m of neritic shelf sandstones and shales which straddle the Devonian–Carboniferous (Mississippian) boundary in North Devon, England. Particular attention was given to the brachiopods and trilobites for their stratigraphic potential. The fauna had been monographed early in the century, and illustrated by lithography. Some revision had been done in the 1950s, but without re-illustration, and there was a good collection in the local natural history museum. What seemed a rather similar fauna of about the same age in Germany had been monographed and photographically illustrated in the late 1930s. Clearly, when the fossils were collected, much work would have to be done on naming correctly and comparing them with German material (and, it was subsequently found, Polish, Russian, Chinese and North African material).

Inland exposures are scattered and few, and the structure of the area complex. The coast section was chosen for study, but it was soon found that it was necessary to map and log the section in detail to appreciate the folding and repetition (Fig. 2.2); this took two weeks. During mapping and logging two tuffs were found which could be used as marker horizons. Potentially fossiliferous beds were noted. It also became clear that the best fossils came from decalcified storm units (§7.2) close to the high water mark. Interbedded shales were somewhat cleaved and fossils distorted, though probably parautochthonous. Suitable blocks of decalcified fine-grained and fossiliferous sandstone were collected from identified beds, though not with known orientation or location within a bed. A problem that was soon noticed was that suitably decalcified material was rapidly exhausted, and that it would take many years for weathering to produce more material. Artificial weathering with hydrochloric acid in the laboratory was not as productive as natural weathering. Particular care was taken to collect part and counterpart, and specimens showing

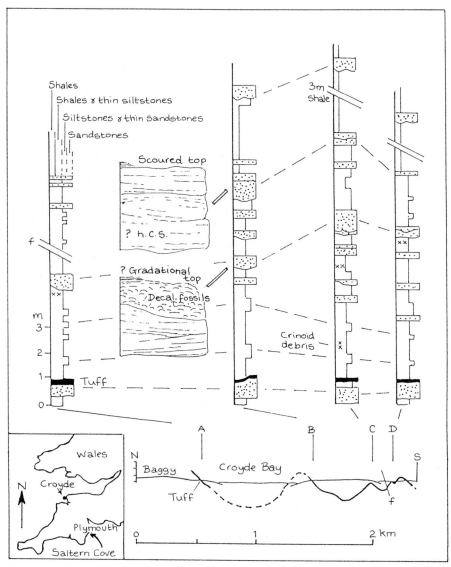

Figure 2.2 Four sections, A–D, with likely correlations in the Upper Devonian of North Devon, England. Field sketches (part) of fossiliferous event beds in section B where they were more decalcified than in the other sections. Note unconventional method of logging heterolithic sediments, which seemed to work and was relatively quick (after Goldring and Bridges, 1973). h.c.s. – hummocky cross-stratification

important morphological features such as brachiopod hinge line and dentition, sculpture, and spines. Some of the best material was found subsequently in blocks dried and split in the laboratory.

The sandstones are typical of storm event sandstones elsewhere in the Phanerozoic, which differ only in the nature of the assemblages and bioturbation. The latter is generally much more intense in post-Palaeozoic storm event beds (Fig. 7.14).

Figure 2.3 View across Croyde Bay, looking northwestwards from B (Fig. 2.2) to Baggy Headland

2.3   An Ordovician hardground (type 6)

Hardgrounds (§7.3.3; Figs 7.17, 7.20, 7.21) are of particular interest to palaeontologists because they have a special story to tell of a colonized, cemented rock surface. In some cases a faunal continuum can be demonstrated, reflecting the gradual change in cohesiveness of the seafloor and its planation by sand under wave action. Palmer and Palmer (1977) gave the first detailed account of an Ordovician hardground and its community structure.

At least 100 m² of hardground was exposed in a road-cut in northeast Iowa, but the surface had been locally damaged by road machinery. An area some 3 m² was found to be undamaged and was marked with a water-soluble marker into 30 cm² areas (Fig. 2.4 and Appendix C), and the encrusting fauna accurately plotted on a half-scale plan. This took three days to accomplish, recording every element greater than 2 mm across. The encrusting and boring biota was composed of 2 types of borings, 3 types of crinoid holdfast and 11 types of encrusting bryozoan. One of the borings was new and, later, required proper description, naming and assignment of type material to the National Museum of Natural History, Washington, DC. In addition a diverse, non-cemented epifauna was collected from the shale immediately above the hardground, which included a sponge, coral, brachiopods, gastropods, other bryozoans, trilobites and a tube-worm. Field identification of the bryozoans raised problems because some had not been described, and because of external homoeomorphy and the small size and immaturity (rather than stunting) of the colonies. Arbitrary 'field' taxa were designated on the basis of growth form and

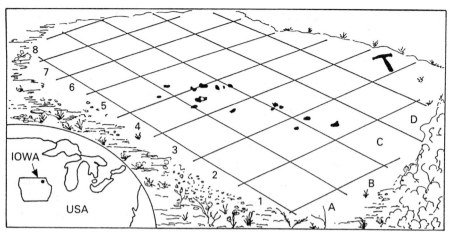

Figure 2.4 Sketch from a photographic illustration of the Ordovician hardground in north-east Iowa analysed by Palmer and Palmer (1977). Blobs give indication of position of colonies depicted in Fig. C1, where quadrat lines are omitted

surface characterization. Where possible, these were subsequently identified with the aid of acetate peels and thin-sections.

The degree of detail was needed because the authors recognized the opportunity to apply work on modern hardground substrate faunas to the Ordovician, and to determine whether controls on larval settlement and intra- and interspecific competition could be recognized and, if so, were similar to those acting today. This aspect was timely because much work was then being done on modern bryozoans and tube worms. Larval settlement on modern rocky surfaces is largely controlled by biotic factors.

In Appendix C, a different statistical method to that originally used by the Palmers is presented. The uniformity of the Ordovician hardground suggests that the strong clumping observed was not due to any heterogeneity, though *Trypanites* (Fig. 6.23) did tend to be associated with low humps. Limited evidence indicated that mechanisms to avoid intra-specific competition were employed by Ordovician bryozoans, and destructive overgrowth between species was also observed. A difficulty in the analysis of interspecific competition in fossils is in establishing that the colonies lived contemporaneously. It was possible to draw up a diagram to show species dominance which indicated that a high degree of interspecific competition already existed in the Ordovician.

Modern hardgrounds are generally fully covered but the Ordovician hardground showed only an average 5 per cent cover. This was attributed to the intensity of erosion destroying the biota, to which Ordovician taxa were much less resistant than those colonizing Mesozoic and younger hardgrounds.

Selected areas and colonies were photographed and specimens from outside the gridded area collected. More work can be carried out on the shale biota and on specific sedimentological aspects; for instance, on the diagenetic history of the hardground using isotope analysis.

2.4   Eocene sands and muds: shelf to shore (types 3, 5 and 9, 10, 12)

The cliffs at Barton-on-Sea, southern England, are one of the world's classic fossiliferous localities (Fig. 2.5); here the Barton and Headon Formations (middle–late Eocene) are exposed. The rocks dip at a low angle to the east and are close to the axis of the Hampshire Tertiary basin (Fig. 2.1 inset). Coastal protection works partly obscure the section. The sequence comprises sands and muds representing depositional environments ranging from open marine to estuarine, lagoonal, fluvial and lacustrine. The diverse fauna and flora have been collected for many years and, in part, described in a range of papers and monographs. No modern comprehensive monograph is available. Plint (1984) gives an account of the sedimentology and palaeoecology of the upper part of the sequence.

Two details are provided from a day field class to the section.

1. The object was to investigate the nature of a facies boundary between two cycles of sedimentation (Fig. 2.6). Plint (1988) has correlated the boundary with the 42.5 Myr global sea level event (Bally *et al.*, 1987). Due to the low dip and long face, the group of students was able to spread out about 2 m apart. The approximate position of the boundary was readily located because of diggings by local collectors hunting for shark teeth. Scraping with a trowel showed glauconitic sandy clay above, and non-glauconitic fine sand and sandy clay below. But bioturbation was so intense that it took some time to locate the actual junction. Confirmation of the level was indicated by occasional pebbles of flint, small logs (generally *Teredo*-bored) and abraded and fragmented shells. The glauconitic sand was seen to be piped downwards with decreasing intensity. Each of the group made a half-scale sketch and the fossils collected were positioned accurately. Some of the fossils were identified at the section using a British Museum handbook but the diversity was found to be far greater than

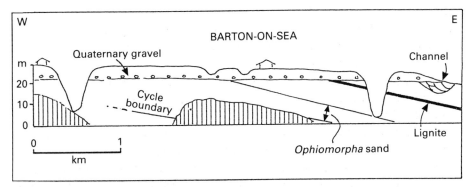

Figure 2.5   Profile of cliffs at Barton-on-Sea, southern England, with site of cycle boundary, lignite and channel referred to in text. Vertical shading indicates where rock is obscured by sea defences. For location in UK, *see* Fig. 2.1

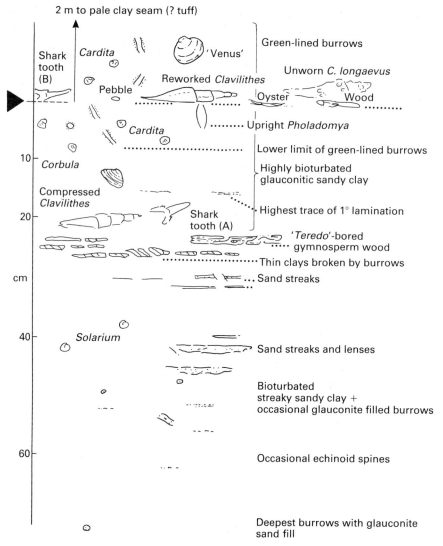

Figure 2.6 Sketch (abbreviated) compiled from field sketches by members of class of cycle (sequence) boundary in Barton Formation, Eocene, at site indicated in Fig. 2.5, with more common elements of biota and notes on lithologies

illustrated. Samples of the sediment were collected from above the boundary in clean plastic bags for laboratory examination of the microbiota and sediment, but sampling below the sequence boundary was difficult because of the bioturbation.

About 1.5 hours was spent on the section, and discussion ensued about the time represented by the omission event and what evidence might be found to indicate events during the non-represented interval. An aspect that seems to have been overlooked by previous workers is the correlation of shells with overlying and underlying strata. The autochthonous burrowers from the erosion surface and small shells and teeth washed into burrows belong to the overlying sediment; while reworked shells from above the boundary belong to the underlying or unrepresented strata. The rare occurrence of brackish water species above the boundary indicates either contemporary allochthonous reworking from a proximal estuarine environment, or that substantial shallowing had taken place during the hiatus. Evidence for erosion having occurred is the presence of worn shells above the boundary with a mode of preservation similar to that otherwise found only below the boundary.

It was noted that many of the shells could be found today in tropical or subtropical locations. This reasoning was supported by the high diversity of the

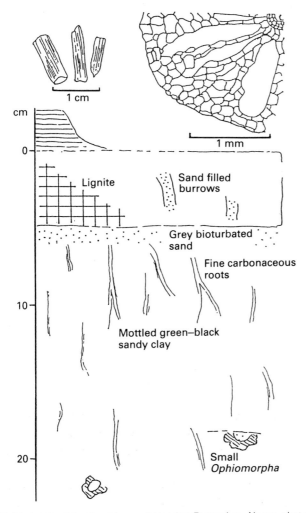

Figure 2.7 Field sketch of lignite at base of Headon Formation. Above, sketches of calcified (anatomical permineralized preservation, §4.2.1) roots of a small aquatic water plant, one with root tip. Thin section, right, shows typical and well-developed air lacunae (after Crane & Plint, 1979)

biota and glauconite, which Hughes (in Hughes and Whitehead, 1987) has shown to be primary.

2. About a kilometre to the east, and almost at the end of the traverse, the lignite marking the top of the next cycle was readily spotted. The objectives here were to determine if the lignite was rooted (Fig. 2.7), to investigate other plant-yielding facies, comparing the yield of different lithologies and, especially, to appreciate Plint's (1984) work in using the biota in facies interpretation. Incidentally, his detailed investigation arose from a freshmens' field-class when he discovered small calcified roots (anatomically preserved; Fig. 2.7) of a water plant a short distance above the lignite (Crane and Plint, 1979).

The lignite was found to be muddy, with roots slender and prominent. Most interest was concentrated on the large channel in the cliff above, especially on the final stage of fill which suggested abandonment. Plint (1984) based his account on his undergraduate project, from which Fig. 2.8 is taken. Making the sketch of the multi-event channel-fill took him some hours of work with a tape and laboriously trowelling out the boundaries. The position of the graphic logs he made, to gain an appreciation of the facies are marked on Fig. 2.8. The partial log (Fig. 2.9) is new and attempts to show the distribution of the plant-yielding layers. Lignitic logs of gymnospermous wood were common but the vertical face made it difficult to collect good leaf material. A bag of shelly silt collected from adjacent beds, which, with gentle washing, later yielded an abundance of shells and *Chara* (green alga) gyrogenites (§4.2.3), almost certainly indicates fresh water conditions.

Although not yielding particularly good material palaeobotanically, when compared with say, Carboniferous sites, this section is important for its record of the major lowstand in sea level (39.5 Myr), and of climatic deterioration during the mid-Caenozoic.

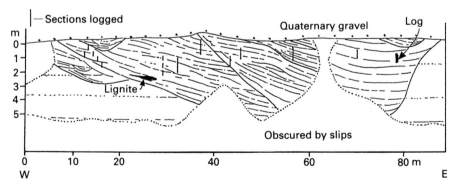

Figure 2.8 Sketch of channel with multi-stage fill, adapted from Plint's undergraduate project report. In the published version (Plint, 1984) areas obscured by slips have been interpreted with full lines. Graphic log illustrated (Fig. 2.9) is from most easterly part of section

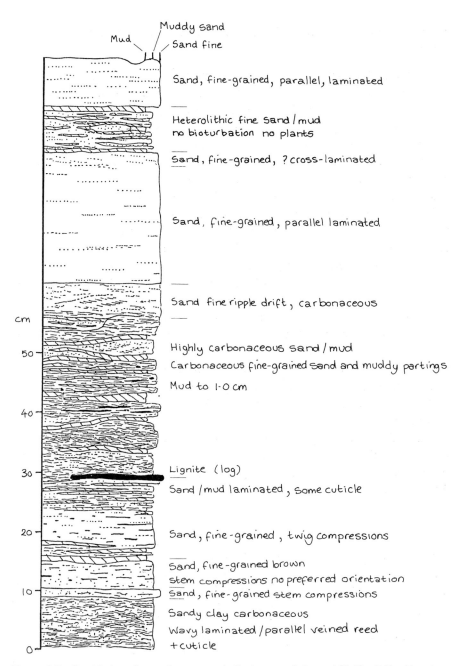

Figure 2.9 Graphic log of part of sequence in final stage of channel fill (Fig. 2.8) with notes on plant fossils

2.5   Condensed pelagic limestones, conodonts, ammonoids and slumps (types 4, 8)

Due to structural complexity, the detailed stratigraphy of South Devon, south-west England, is still largely unresolved. Biostratigraphers must look outside the local area, to geologically less complex regions, such as Central Europe, where the biotic sequence can be readily established, in order to unravel events in Devon.

Such an approach was adopted by Maurice Tucker, during the course of his postgraduate work, whilst investigating pelagic carbonate sediments. In conjunction with Peter Van Straaten, who had recently mapped an area of Devonian rocks in Germany, applying conodont and ostracode biostratigraphy, Tucker re-examined the Saltern Cove section (Figs 2.10, 2.11). This is a classic goniatite locality, containing hematized goniatites, extracted from shales, belonging to the lower part of the Upper Devonian (Zone 1δ). In order to determine the sequence of events, Van Straaten and Tucker had to establish both the age of the associated underlying and overlying shales, and the age of the blocks of limestone found in the goniatite 'bed'. Sampling was not done by any statistical method, but by the practicality of sampling the various lithologies and marking collecting points accurately on the sketch. Conodonts were obtained by breaking open the shales and searching for specimens with a hand lens. Most were in small 'reduction' centres. Later, conodonts were extracted by formic acid digestion from 10–20 cm limestone blocks.

The results showed that the age of the goniatite-bearing unit was not as simple as had been thought. The underlying and overlying shales were much younger, as also were the clasts associated with the goniatite-bearing shales. Van Straaten and Tucker (1972) offered a convincing reinterpretation of the sequence indicating deposition, reworking and *en bloc* transportation, probably from a Devonian oceanic rise, into a basin.

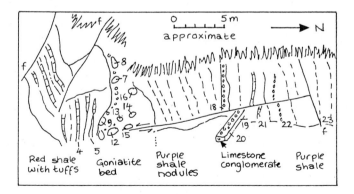

Figure 2.10 Sketch, simplified from that published, of Saltern Cove, South Devon, viewed obliquely with some of the sampling points (for location see inset to Fig. 2.2)

Figure 2.11 Photograph of cliffs at Saltern Cove, South Devon, in Upper Devonian sediments and depicted in sketch Fig. 2.10

The methods used are typical of those employed by structural geologists and sedimentologists working on tectonically complex sequences.

2.6 Pelagic bituminous mudrock: aerobic, dysaerobic or anaerobic? (types 7(5))

Undertaking an investigation of the type section of the Upper Jurassic Kimmeridge Clay, Aigner (1980) expected to encounter bituminous and finely laminated mudrock. The extensive cliff sections do display such facies; but on the broad rock platform, and between boulders, a number of bedding parallel features (§7.1) such as strings of phosphatic coprolites, imbricated ammonites, shell pavements with occasional shell-filled gutter casts, sharp-based graded units, and oyster encrusted ammonites, suggest that the depositional environment was by no means as continuously quiet as had been postulated. Rather, appreciable turbulence, probably introducing oxygenated water, is envisaged.

The work by Aigner (1980) thus emphasizes that these sediments, some 280 m thick, could not have been a continuous hydrocarbon generating system, and that an appreciation of this could lead to a better understanding of the source rock.

Wignall (1989), as part of an extensive postgraduate study of the palaeoecology of the Kimmeridge Clay, has been able to show (Fig. 2.12) that the shell pavements probably represent parautochthonous accumulations due to storm reworking.

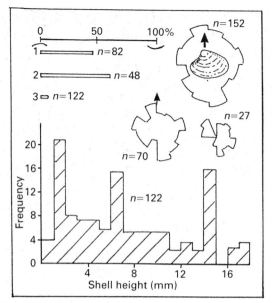

Figure 2.12 Shell taphonomy in the Kimmeridge Clay, adapted from Wignall (1989). (Upper left) proportion of convex-up to concave-up valves from three shell pavements. (Upper right) rose diagrams of orientations of disarticulated bivalves on bedding surfaces; two were from fallen blocks and the third was collected in place. (Lower) size – frequency histogram of *Protocardia* valve height on a single bedding surface, showing three peaks, probably indicating three successive recruitments of an opportunist

2.7   Actuopalaeontology: the palaeontology of the present

'A sight of the sea floor is worth a hundred bedding planes'

This section briefly describes what can be achieved by traverses across modern coastal sections: a Californian sub-tropical beach, and three contrasting, temperate sites in South Wales. Each afford an opportunity to study how and where fossils might have once lived, and how the modern shells might eventually be buried to become the fossils of the future.

Three lines of enquiry should be followed at modern sites:

1. *Questions about the present*: how animals and plants live, interact, grow, reproduce; their tolerance to salinity, turbulence; feeding modes, mobility; causes of death; how organisms are incorporated into the sediment and what proportion is preservable.
2. *Questions about the future*: the effects of relative rise/fall in sea level, increase in sediment input, a hurricane, volcanic eruption or introduction of new taxa or of others becoming extinct. The effects of channel migration or pollution.

Figure 2.13 Microatolls formed by the coral *Porites* in shallow water at Heron Island, Queensland, Australia. Upward growth is inhibited by the air–water interface. Scale bar = 0.5 m

3. *Questions relating back to the ancient sediment*: how do observations on modern sediment differ from those made on ancient sediments; what changes in ecological factors have taken place? How have evolutionary changes affected environments?

2.7.1   A subtropical beach in southern California (Figs 2.14, 2.15)

Bahia la Choya, a few kilometres northwest of Puerto Penasco (latitude 31 °N), is at the northern end of the Gulf of California (a rifted basin which opened 3.5–4.5 Mya). The climate is arid and the summer surface water temperature is 30–32 °C (10–14 °C winter). Salinities are slightly raised in shallow coastal waters and the tidal range, up to 9 m, is associated with rapid flooding. There is marked intertidal zonation of the biota.

A 4 hour walk across this relatively sheltered site, from the outer flats to the salt marsh, demonstrates the zonation, associated bedforms and biogenic structures.

The outerflats are composed of medium- to coarse-grained sand, with flood-oriented sand waves and superimposed smaller ripples, also ebb oriented, or with respect to late stage drainage. Sand wave troughs are often shelly with washed-in sand dollars (*Encope*) and disarticulated bivalves. Live animals are relatively rare in this high energy environment. Depressions excavated by rays searching for bivalves are common and, locally, there are patches of vertical, shell-lined burrows (*Diopatra*). The surface trails of predatory *Natica* have no

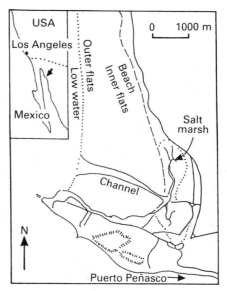

Figure 2.14 Sketch map of Bahia la Cholla, 7 km northwest of Puerto Penasco, northern Gulf of California showing major habitats. Course of channel schematic

chance of being preserved as fossils, but the morphology of the trails gives a clue to the way fossil tracks, probably made by gastropods, were formed.

In contrast, the inner flats are characterized by smaller grain size and essentially small-scale, ebb oriented, though wind modified, ripples. There is a low topography of slightly raised areas of algal-bound, fine sand, broken by depressions of pellet-rich, soupy fine sand. The live fauna includes deeply burrowing razor shells, *Tagelus*, and the scavenging gastropod *Nassarius*. Although displaying little surface expression, the extensive burrow systems of the ghost shrimp *Callianassa* can be found by digging. If time allows, it is useful to discuss how closely the burrows resemble fossil burrows such as *Thalassinoides* (Fig. 6.23G) (but note that the species of callianassid shown here does not line its burrow wall). There is a concentration of shell material at a somewhat variable depth resulting from sediment-sorting activities of deposit feeders. When digging on the muddy inner flats the boundary between oxic and anoxic sediment can be seen very close (only 1 mm) to the surface in contrast to the sandy channel sands where it is several decimetres down.

Aspects that may be usefully discussed are the variation in shell destruction in the different environments: physical and biological destruction being highest in the outer flats and channel, whereas chemical destruction, scarcely evident on the outer sands, becomes an important aspect in the preservation of salt marsh shells. In the uplifted Pleistocene deposits around the bay only impressions or chalky remains of formerly aragonitic shells survive. An aspect that is not obvious during the traverse is the role of reworking. Some of the more robust shells have been radio-carbon dated as greater than 3000 yBP indicating extensive time-averaging of the biota and loss of the more delicate shells.

A more complete guide to the area can be found in Flessa (1987).

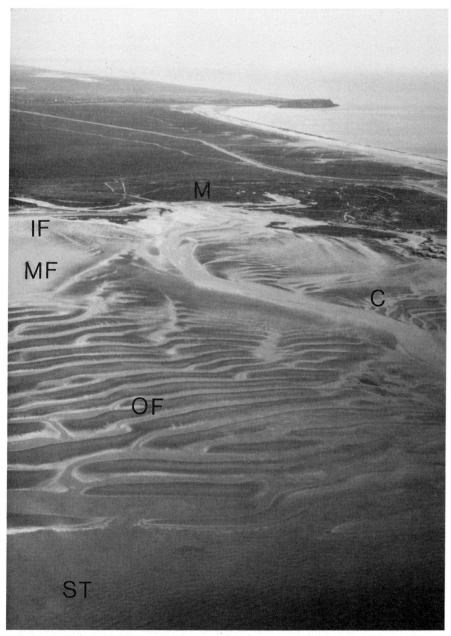

Figure 2.15 Oblique aerial photograph of Bahia la Cholla.
M: marsh; IF: inner flat; MF: mid flat; OF: outer flat; C: channel; ST: subtidal

2.7.2 Three contrasting, temperate intertidal sites in South Wales

Gower in South Wales (51° 30' N) is on the north side of the Bristol Channel. The tidal range is up to 8.6 m. Salinities on the open beaches are normal and offshore bottom water temperature is 16.5 °C (August) and 7.5 °C (February).

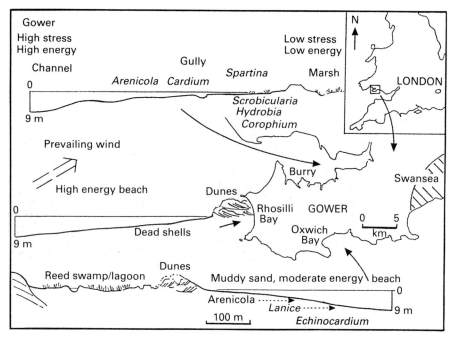

Figure 2.16 Distribution of biota at three beaches in South Wales: a high energy beach (Rhossili, Fig. 2.17), facing into prevailing wind with fauna living only at low water; a medium energy beach (Oxwich Bay) with *Arenicola* flats, dune bar and back-swamp-lagoon, and an estuary (Burry) with tidal channel and sand and mud flats (Fig. 2.18). In the estuary, energy (stress) decreases towards the mud zone at high water, where the shoreline is prograding due to colonization by *Spartina townsendii*

A visit to three contrasting beaches was made with low-water about 1300 h in order to demonstrate the differences in sediment, sedimentary structures and biota, and ecological controls on the biota and the taphonomy.

Rhossili beach (Figs 2.16, 2.17) is exposed to the prevailing southwest wind and Atlantic storms. At the high water mark there is a line of debris left by the receding spring tide: shells, arthropod carapaces, drift wood (with attached epiplankton), seaweed, dogfish and whelk egg cases, pebbles, fish remains, bits of sponge, bryozoans and the inevitable plastic and other garbage. Many of the shells will have come from just offshore (*Donax*) or from the nearby rocky coast (*Mytilus, Buccinum, Natica, Littorina*). Occasional land shells have come from the dunes behind the beach. The sand of the upper shore is smooth except for primary current lineation and traces of antidunes. At times of high wave intensity the middle shore is thrown into sand waves which are only incompletely levelled as the tide recedes. Digging shows no sign of life in the sand and, close to high water, displays undisturbed beach lamination. Towards low tide level tubes of the sandmason *Lanice* are conspicuous, and if the tide is very low one can spot the small openings of the rapidly burrowing razor shells.

In contrast, the beach facing to the southeast at Oxwich has a different aspect. The lower beach is covered by ripples, mostly somewhat degraded, and lugworm

Figure 2.17 Rhossili beach, looking north, at low water on a calm day

castings (Fig. 2.20). Digging reveals no lamination. Towards low water mark *Lanice* (Fig. 2.21) may be abundant, but areas where this is absent may show the small inhalant opening of the potato urchin (*Echinocardium*). With care it is possible to see an emerging tube-foot maintaining the tube. Digging will not reveal the 'orange peel-like' layers formed by the animal's forward progression, these only show up in fossil burrows by diagenetic enhancement (*see* §4.4).

At Penclawdd (Fig. 2.18) the salt marsh grades into mud and sand flats which are traversed by dendritic tidal channels (contrast with the drainage system formed in tropical mangrove environments, Fig. 7.1). Here, energy increases away from the shoreline and the upper flats are very muddy. Grazing *Hydrobia*

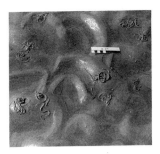

Figure 2.20 Penclawdd, lower sand flats with sand ripples and castings of *Arenicola marina* (lug-worm). Scale bar = 1 cm

Figure 2.18 Penclawdd, Burry, looking northwest. Clumps of *Spartina townsendi* colonizing upper shore muds

Figure 2.21 *Lanice* (sand-mason) tubes being eroded in small ebb-tide channel, Belmullet, Mayo, Eire

Figure 2.22 *Mya* shells being eroded from sediment at outside of meander bend of small channel, Penclawdd, Burry

Figure 2.19 Burry estuary, tidal channel with imbricated valves of *Cerastoderma* and *Scrobicularia* (looking downstream)

are present in high numbers. There are two types of conspicuous trace: paired openings of the amphipod *Corophium* with its *Diplocraterion*-like burrow (*see* Fig. 7.18), and the star-trace above *Scrobicularia*, an infaunal deposit-feeding bivalve. On the sandier flats, scraping the surface with a fork shows many cockles (*Cerastoderma*). Annual growth and disturbance rings are clear. The small tidal channels are often floored with a layer of cockle shells (Fig. 2.19) and it is easier to walk along the channel than across the mudflat. Towards the main channel evidence of storm degradation is common on the sandflats (Fig. 2.22). Ripple troughs often display abundant echinoid spines washed into the estuary.

Discussion might concentrate on the main ecological controls operating on the biota at each site, especially on the role of the energy regime.

References

AIGNER T 1980 Biofabrics and stratinomy of the Lower Kimmeridge Clay (Upper Jurassic, Dorset, England). *Neues Jahrbuch für Geologie und Paläontologie, Abhandlungen* **159**: 324–38

BALLY A W (ed) 1987 *Atlas of Seismic Stratigraphy*, Vol. 1, American Association of Petroleum Geologists, Studies in Geology, No. 27

CRANE P R, PLINT A G 1979 Calcified angiosperm roots from the Upper Eocene of southern England. *Annals of Botany* **44**: 107–12

FLESSA K W (ed) 1987 *Paleoecology and Taphonomy of Recent to Pleistocene Intertidal Deposits, Gulf of California*. The Paleontological Society Special Publication 2, 237 pp (Keys to shells of Bahia la Choya and collections of papers on taphonomic and ecological aspects of biota)

GOLDRING R, BRIDGES P 1973 Sublittoral sheet sandstones. *Journal of Sedimentary Petrology* **43**: 736–47

HUGHES A D, WHITEHEAD D 1987 Glauconitization of detrital silica substrates in the Barton Formation (Upper Eocene) of the Hampshire Basin, southern England. *Sedimentology* **34**: 825–35

PALMER T J, PALMER C D 1977 Faunal distribution and colonization strategy in a Middle Ordovician hardground community. *Lethaia* **10**: 179–99

PLINT A G 1984 A regressive coastal sequence from the Upper Eocene of Hampshire, southern England. *Sedimentology* **31**: 213–25

PLINT A G 1988 Global eustacy and the Eocene sequence in the Hampshire Basin, England. *Basin Research* **1**: 11–22

VAN STRAATEN P, TUCKER M E 1972 The Upper Devonian Saltern Cove goniatite bed is an intraformational slump. *Palaeontology* **15**: 430–8

WIGNALL P B 1989 Sedimentary dynamics of the Kimmeridge Clay tempests and earthquakes. *Journal of the Geological Society, London* **146**: 273–84

Texts on Sea Shore Life

(*See also* texts in Chapter 4 *Taphonomy* and Chapter 6 *Ecology*)

ABBOTT R T, DANCE S P 1982 *Compendium of Seashells* E P Dutton Inc

BARRETT J H, JONGE C M 1958 *Collins Pocket Guide to the Seashore* Collins, 272 pp (For northwest European shores)

CAMPBELL A C 1976 *The Seashore and Shallow Seas of Britain and Europe* Country Life Guides, 320 pp

KOZLOFF E N 1983 *Seashore Life of the Northern Pacific Coast* University of Washington Press, 370 pp

KOZLOFF E N 1987 *Marine Invertebrates of the Pacific Northwest* University of Washington Press, 511 pp

MORRIS R H, ABBOTT D P, HADERLIE E C 1980 *Intertidal Invertebrates of California* Stanford University Press, 690 pp

MORTON J, MILLER M 1968 *The New Zealand Seashore* Collins, 638 pp

REINECK H-E 1970 *Das Watt* Kramer, 142 pp (For German Watten Sea)

3

Field strategies

'If you know what it is you are searching for then you will find it'

The aim of this chapter is to discuss essential field strategies. Tools, techniques and tactics are summarized in Appendix A, with additional information in Chapters 6, 7 and 8. Assessing the literature and making a reconnaissance are often tedious but always well worth the trouble, especially since they enable one to formulate more clearly the objectives of the field study, and to plan the most economical use of the time available in the field.

A section on stratification is included here because the bed is the basic sedimentary unit, and any fossil that is found *in situ* has to be related to the underlying and overlying bedding surfaces. Changes at bed contacts reflect change in sedimentary environment and become the most conspicuous and, indeed, most important lines on the sedimentary log. An appreciation of bedding is therefore essential in logging outcrops and cores.

Most specimens collected in the field are, in reality, samples that may be subsequently used to illustrate an aspect of the facies or biota. If the aims of any investigation are carefully defined then a sampling plan can be formulated. The field strategy that is required for buildups is rather different from that required for stratiform fossiliferous occurrences. For both there is no short cut to the basic mapping of the site, and accurately recording the exact location of observations and specimens collected.

Basic field mapping and surveying techniques are covered by Barnes (1990), Butler & Bell (1988), Compton (1985) and Moseley (1981).

3.1 Previous work

Although there is something to be said for going into the field with an open mind unprejudiced by the findings of previous workers, afterwards most of us generally regret having done so. The objective in making a literature survey is to establish the 'state of the art', the distribution and age of the rocks, the structure of the area and current problems. One should get as good an appreciation as possible of the biota likely to be encountered, so that the taxa can be readily recognized.

Begin with a regional textbook and follow with more detailed reports, 'Survey' memoirs and monographs. There is obviously a limit to the amount of literature that can be taken into the field but glean information on locality lists and note preservation modes from the illustrations. It may be necessary to refer to books or papers borrowed from distant libraries, and some may need translation. Inspection of museum collections can prove valuable. Even more useful is to discuss plans with someone already familiar with the area. These aspects can take a lot of time: hence the need for ample preparation. When in the field contact local museums and geologists.

Allow for the possibility that fossil names in older literature may have been subsequently revised, and also, that illustrations in pre-photography days are not always as accurate as might be expected. Be prepared for cases of misidentification in older fossil lists and do not expect that collecting sites of a century ago will still be available. With older literature note that many new fields have since been researched. For instance, while faunal lists in a century-old paper may be detailed, it is unlikely that there will be substantial reference to the palaeoecology or sedimentology.

Be sceptical of claims in the literature that sediments are unfossiliferous, as they may abound with trace fossils. At most, unfossiliferous only means that no fossils have yet been found.

3.2 Reconnaissance

The object of a reconnaissance is to gain an appreciation of an outcrop or area sufficient to know where the best sections are situated, their limits and weathering profiles, the general accessibility and local problems (including obtaining permission to enter land), what tools, recording and photographic equipment are required and whether or not special techniques, such as peel making, are likely to be needed. Also get a good idea of the time required to carry out the objectives.

If good loose material is spotted, collect or flag it, since it may not be readily relocated, but do not spend time on detailed investigations. Nevertheless, this is the time to carry out carefully considered pilot studies. When the reconnaissance is completed re-assess the objectives and the feasibility of completing them, and formulate as detailed a programme of work as possible, include a listing of the jobs required, putting these in order of priority.

3.3 Stratification, bedding and cyclic sedimentation

The bed is the basic sedimentary unit. It can be recognized in most sedimentary facies, including many autochthonous buildups. Figure 3.1 attempts to

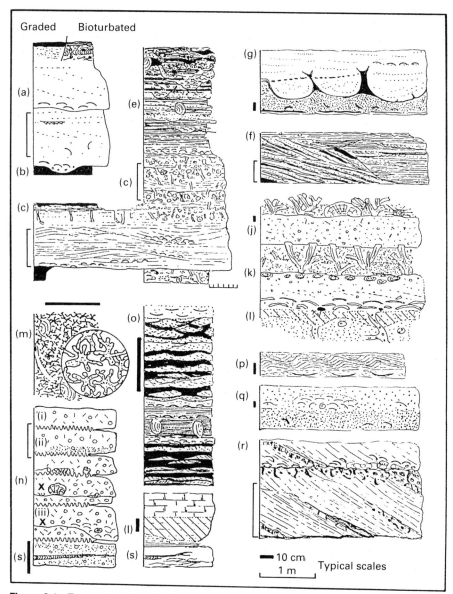

Figure 3.1 Types of bedding (for further explanation see text)

portray in two dimensions the principal types of bedding that are generally associated with body and trace fossils. Note how difficult it may be to fix the lower as well as the upper boundary of many types of bed, in much the same way as the substrate surface is often more or less transitional between water and sediment.

The term bed is frequently applied to sedimentary units greater than 1 cm in thickness, below which lamina is applied. This is an arbitrary definition.

Lamination is present within a bed and results from the physical processes operating during the formation of a bed. Do not refer to a thickness of sediment with lenticular or heterolithic bedding as a bed, because each alternation was formed under a different sedimentary regime. But a unit of climbing ripple lamination may constitute a bed. For different thicknesses of bed the most convenient method of designation is as mm bedding, or mm–cm, cm, dm, etc. Also consider carefully before using bed for an interval of mudrock. Event bed is commonly used for a storm, turbiditic unit, or tempestite marking a single rapidly deposited sedimentary unit (§7.1).

There are several types of bed

(a) *Turbidites* Trace fossils on the sole of a bed indicate pre-turbidite and deeply penetrating, post-turbidite organisms (§7.3.1). Trace fossils on the upper surface indicate the shallow-penetrating post-turbidite suite possibly introduced with the turbidite flow. Within turbidites, which can be composite, body fossils tend to be scattered, but with preferred orientation and often convex surface down. Shells may be relatively uncompressed.

(b) *Inter-turbidite mudrock* is generally sparsely fossiliferous, but can yield important material. Water depth is reflected in preservation of calcareous skeletons (§6.2.3.1). Release of fossils often depends on cleavage relationships and weathering.

(c) *Storm beds and tuffs* Trace fossils on soles and upper surfaces are less strongly differentiated compared with turbidites. Body fossils occur in lags, concentrated in gutter casts, or in lenses associated with cross-stratification (including hummocky and swaley cross-stratification with, respectively, convex-up and convex-down intra-bed discontinuities). Fossil fragments are often scattered through a bed. Amalgamation is common, and each unit (bed) may be associated with a different biota. Event beds may cover (smother) a biota (*see also* Fig. 7.9).

(d) *Graded muddy sandstones or siltstones* (not figured) (*marine shelf*) are generally poorly fossiliferous.

(e) *Heterolithic stratification*, alternating mud and sand grade sediment, is a term often applied to irregular alternations of mm–cm thickness. Where weathering allows, sandstone soles are often rich in trace fossils e.g. *Cruziana*. Look for shelly partings, lenses and gutters. The shelly biota is often rich and diverse and often autochthonous.

(f) For palaeobiological purposes *point bar sedimentation* can be included with heterolithic stratification. The laterally-accreted alternations of thin mudstones and siltstones of fluviatile and estuarine facies are often rich in plant fossils.

(g) *Load-cast bedding* (often with evidence for reactivation when in shallow-water facies, especially estuarine facies) is generally poorly fossiliferous. If present, shells are distributed as in (c).

(h) *Convolute lamination* (not figured) is generally poorly fossiliferous.

(i) *Paleosols* (not figured) formed under various climatic conditions, mark emersion events and may be useful markers. Evidence of roots (± carbonaceous remains) is unusual. A molluscan fauna may be present that was associated with the soil, or introduced with subsequent submersion (*see also* §1.3).

(j) *Event bed* of bioclastic grains truncating and smothering coral growth.

(k) *Firmgrounds and hardgrounds* are associated with omission in sedimentation and recognized by colonization by encrusters and borers or burrowers. A mineralized crust is typical. As depicted, coral growth follows a slightly irregular firmground of oncolitic packstone (*see also* Fig. 7.17).

(l) *Under-'beds'* form by the underlying lithology being cemented to the overlying limestone, thus obscuring the sole (and associated trace fossils and sedimentary structures), and generally making it difficult to extract fossils from the underlying sand (right) or mud (below centre).

(m) *Growth bedding* has not been recognized in ancient sediments. Today it is forming where *in situ* growth is taking place within a unit of free-living calcareous algae. *Lithothamnion* is depicted.

(n) *Pseudo-bedding* (Simpson, 1985) is produced by pressure-dissolution, stylolite-controlled layering, in generally homogeneous limestones: typically packstones to grainstones.

   (i) *Unverifiable pseudo-bedding* where a stylolite parts identical lithologies.

   (ii) *Exaggerated true bedding* where a stylolite separates different lithologies.

   (iii) *Verifiable pseudo-bedding* where true bedding (x) can be identified (often cryptic) between stylolite surfaces. At outcrop stylolites often weather back as prominent surfaces, or splay into anastomozing feathery lines; a situation typical of fine-grained carbonates such as Chalk (opposite – (s)).

(o) Sequences of *cross-bedding* with mud flasers (wisps) transitional to wavy bedding to lenticular bedding (lenses of sand). Trace fossils are often distinct on soles but shelly biota is generally sparse. Centre, sequence of parallel mm thick beds, typically of silt grade. Undertracks may be present. Muddy laminae may be associated with shell stringers and pavements. Herring-bone cross-stratification is a good indicator of tidal environments but check whether opposing foreset dips are only apparent. Couplets (typically mm sands separated by a thin mud: see base of figure) are also indicative of a tidal regime.

(p) *Ripple-drift lamination* is generally poorly fossiliferous, even for microfossils because of poor development of ripple troughs.

(q) Diagenetically preserved *'cluster' of shells* within an early formed concretion. Subsequent loss of shells in surrounding sediment.

(r) *Cross-stratification* in tidal sand-grade sediments. Trace fossils occur at the base of a bed and rise up foresets corresponding with phases of low sedimentation (often muddy laminae). Shelly fossils, bones and wood are mostly concentrated along bottom sets.

(s) *False indications of bedding* (particularly in cores) may be due to laminar trace fossils such as *Zoophycos*, or even, bedding-parallel simple burrows.

(t) *Unconformity* (Fig. 8.3).

(u) *Biogenic stratification* (not figured) (Meldahl, 1987) and *biogenic graded-bedding* (Trewin & Welsh, 1976) have only been recognized in modern sediments. Both result from the activity of infaunal organisms. In biogenic stratification fines are eliminated by the activity of polychaetes and arthropods. The lower limit of their activity is relatively sharp but never as sharp as with hydraulic processes. Biogenic graded-bedding results from size sorting of grains, especially shells, with the concentration of larger particles in the lower region of the animal's activity.

(v) Useful indications of stratification (not figured) are *layers of concretions* and *diagenetic structures* such as burrow filled flints and cherts. The latter relate to an original sedimentary surface well above the actual layer. Diagenetic bedding can be particularly difficult to prove. The origin of thin stratiform concretions is generally clear and a termination to the layer as it passes into normal sediment demonstrates its origin. Commonly diagenesis has followed primary bedding.

Cyclic sedimentation (rhythms when asymmetric) with regular repetition of facies or sequences of facies ranging from mm to many m in thickness can be due to a variety of causes: (1) local controls: epiorogeny, local faulting, storms, diagenesis (many limestone–mud alternations), the casual migration of a river over a floodplain (giving fining-upward fluviatile cycles) which are periodic or aperiodic; (2) global controls: plate margin activity, ocean-ridge activity, glacial control, effecting eustatic change; and (3) orbital control with periodicities of 19, 23, 41 Ka upwards affecting particularly climate (and hence thermal expansion–contraction of oceans). Description is dependent on careful field observation.

3.4 Graphic logs (Fig. 1.2)

A graphic log, compiled using similar techniques employed in the production of a large scale map, provides the best appreciation of a section. It enables evaluation of the facies, and provides a 'ladder' on which to position the fossils observed and collected. A good log acts as a summary diagram of all the information about a section. It must be readable. For palaeontological purposes realistic and annotated logs are the most useful. Stylized logs are constrained by the symbols available and can give a misleading impression. For example, if the top of a bed grades imperceptibly from sand–mud, then no filter-feeding organism could have colonized the fine-grained substrate. Drawing a realistic log makes one think of the implications of every line.

Decide whether it is necessary to log the whole sequence, and to do so at the same detail. Or, is just a representative log required? This is more difficult because it assumes one knows what facies are to be selected. Decide whether the log is to be essentially one-dimensional (i.e. borehole, stream section), or two-dimensional, where an extensive face exhibits lateral variation.

Logging strategy:

1. Inspect section, clean up loose material, remove vegetation to expose important boundaries. Mark important features. It may be necessary to make a grid using string, chalk or pegs if it is intended to record lateral changes.
2. Scale: on a log sheet or in a notebook it is not possible to record a unit in less than 2.0 mm. Thus, if the smallest unit to be recorded is 50 mm then the scale required will be 1 mm on the log equivalent to 25 mm at the outcrop or core. Details can always be shown at a larger scale.

$$\text{Logging scale (LS)} = \frac{\text{thickness of smallest feature to be measured}}{2.0} \text{ (mm)} \quad \text{e.g.} \quad \frac{50}{2} = 25 \text{ mm}$$

3. Prepare notebook or logging sheets with columns headed according to the information required (below). Keep a column for notes.
4. Recording: if possible work upwards (stratigraphically) rather than downwards. This way it is easier to gain an appreciation of the nature of bed and facies contacts. At the outcrop, first mark in the weathering profile. In siliciclastic sediments this generally corresponds to lithology and grain size, but in carbonate sediment dissolution seams may be present and autochthonous (reefal) units may weather back more than a micrite. Lithology can then be added against the weathering profile or as a separate column, and grain size as a further column. In well-sorted sands, grain size estimation does not present any problem, but in poorly-sorted sediments, estimate the mode and then add a point or points to indicate range of grain size present. In carbonate sediments, more emphasis is placed on sorting. Plot wackestone, packstone and grainstone. Add bedding and other sedimentary structures. Lightly colouring a log aids visual impression (e.g. clay, light blue; silt, grey; sand, yellow; gravel, brown).

Distribution of fossils on logs presents problems. Rows of symbols are quite unsatisfactory, though some may be added to indicate preservation mode and completeness. Frequency (§6.2.3.7) may be indicated by arranging the columns with frequency decreasing left to right or by line thickness.

5. Photograph the logged section and mark the areas photographed on the log with the photo number.
6. Indicate sampling points on log by sample numbers.
7. Go over log checking entries, make a preliminary analysis of the facies and note the presence of cycles and sequences.

Illustrations of other types of logging sheets are given in Graham (1988).

3.5 Sampling

What size of sample is needed, and at what interval the samples should be taken, are common questions in the field. Particular consideration, however, should be given to:

1. The sampling interval at which the products of the biological processes can best be identified.
2. The sample size necessary to ensure that it will yield data to identify the pertinent patterns of biological or taphonomic processes and attributes (community identification, biostratigraphy, morphological variation).

Although not essential for the field, it is often useful to be aware of the statistical tests that can be applied to the samples, especially if the number of samples is less than 30 (Appendix C).

The sample interval

Determine:

(a) The process and its rate of operation. Biological processes that are of most concern are: growth, evolution (for biostratigraphy) and ecological change. Many biological processes are too rapid or are unrecordable (e.g. angiosperm fertilization).
(b) The net sedimentation rate for each facies.
(c) Event beds and penecontemporaneous erosion/non-deposition. Event beds normally need be sampled only 'once' regardless of thickness. But, ascertain just how the biota is distributed through an event bed.
(d) The sequence of facies involved.

In most cases (a) and (b) can be no more than estimates. Some values and their ranges are:

Speciation may operate over	10^5–10^6 yr (for biostratigraphic fossils)
Ecological succession	10^0–10^3 yr

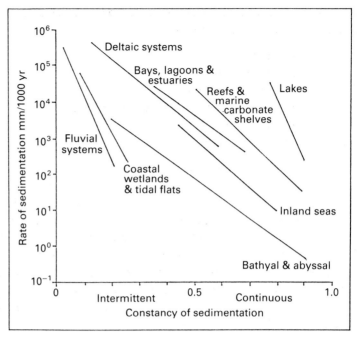

Figure 3.2 Relationship between rate of sedimentation (shown as mm/1000 yr) in various modern sedimentary environments, and constancy of sedimentation i.e. fraction of time during which sediment actively accumulates (adapted from Schindel, 1980)

Habitat destruction	instant–10^3 yr
Regional extinction	10^3–10^4 yr

Sedimentation rates for various environments are shown in Fig. 3.2 with a range of values for each to allow for aggradation and degradation. But use the figures with caution.

The sample size

Determine the frequency of a taxon per unit volume/weight of sediment, or frequency of the taphonomic effect, for instance, valve dissociation. For bulk samples of macrofossils it is best to determine a sampling curve (Appendix C–6) corresponding to size of sample and number of species collected. For instance, in the Silurian of Wales, increase in species number fell off at a total of 200 individuals, so that any sample should contain at least 200 individuals. A 'bucketful' should provide sufficient numbers from neritic sandy mud, but search along the outcrop for rare species that may be important for reconstructing upper trophic levels. In many instances one will be searching purposefully for particular modes of preservation, or for stratigraphically useful fossils. No statistical inferences can be made from such collections about distributions or trends. The axiom of statistical sampling is that it should be free of bias. At exposure, consider the following:

1. What is to be determined?
2. Why is it not feasible to measure everything?
3. What degree of precision is required?
4. To what extent will the sampling pattern be controlled by lack (gaps) of exposure, accessibility, differential weathering?
5. What sampling method is required?

For *trends* over an area or along a transect, for instance, decrease in size or frequency of a taxon, changes in diversity, adopt a systematic or stratified random sampling scheme (Fig. C5). Systematic sampling is useful when sampling points are to be spread evenly, and it is a simple method. If systematically sampling, select the first sample position randomly. For *distributions*, for example, orientation, or the amount of cover of different taxa on a bedding surface, use a simple random sampling scheme. If it is intended to sample a particular facies, for instance only the limestones of a limestone – shale rhythmic sequence, then take samples at regular or random intervals as the case may be, above the base of successive limestones. When sampling to test similarity or difference adopt the same sampling scheme for each unit, for example whether the taphonomic makeup of two beds is similar. A set of random numbers (Table 3.1) can be used, or a pair of dice thrown, to generate random intervals.

Table 3.1 A table of 100 random numbers. Use single numbers as spacings (e.g. cm, mm) or pairs of numbers as co-ordinates

72	20	09	65	77	70	94	34	25	85	96	11	78	07	30	61	03	04	63	92
06	01	76	27	90	68	71	38	29	23	13	59	93	52	83	44	53	37	91	21
57	79	86	47	80	35	81	05	62	50	33	15	73	39	42	60	55	19	56	87
48	40	64	67	16	12	98	32	84	69	22	17	24	28	02	97	66	45	41	43
58	00	54	46	82	74	36	49	31	14	89	08	18	95	88	99	10	75	51	26

If a quadrat is used then its size should be related to the size of the fossils (plant fragments or shells). Size can be determined with a pilot study, by taking quadrats of increasing magnitude and plotting the increase in number of taxa recorded. Quadrats may be marked out with chalk, or staked out with skewers or tent pegs. A quadrat with a 0.5 m side will be suitable for most situations. The quadrat frame is covered with a clear plastic sheet on which 100 randomly distributed points are marked. Record the identity of each fossil at each point. The data can be analysed either, on a percentage basis, or by ranking (e.g. Domin scale where + = a single individual, 1 = 1–2 individuals; 2 = less than 1%; 3 = 1–4%; 4 = 5–10%; 5 = 11–25%; 6 = 26–33%; 7 = 34–50%; 8 = 51–75%; 9 = 76–90%; 10 = 91–100%).

3.5.1 Microfossils and palynofacies

For *microfossil* and *palynofacies* analysis, only material that is *in situ* should be used for stratigraphic purposes. It is vital that the sample is unweathered and clean and, for pollen and spores, has not been exposed to the air. Use new polythene sample bags, or tubes, with a closure. The distribution of micro-fossil groups in different lithologies is shown in Table 3.2. Lithologies are given in general terms. For instance, foraminifera will only be found in marine sandstones, unless derived. Metamorphosed sediments are omitted but may yield rare instances. Lenses of limestone or chert between lava pillows are important sources. Take *spot* samples at regular intervals, or over a regular number of events (allow for amalgamation of turbidites; Fig. 7.9), or *'channel'* samples, where sampling is continuous over a metre or two (thus reducing the possibility of accidentally sampling unfossiliferous intervals). Sample size will depend on the nature of the sediment. As a general guide for ostracodes and foraminifera in fossiliferous mudstones etc., 200–500 g will yield sufficient specimens. For coccoliths 30 gm is sufficient. But, for sampling highly 'diluted' sediments, for instance, conodonts in an encrinite, several kilograms may be needed for later acid digestion. Note that in bioturbated sediment (including most chalks), there may be no point in sampling at closer intervals than 0.5 m for stratigraphic purposes because of the redistribution of the biota. Samples for later palynological analysis (palynofacies) may be important for correlation and environmental interpretation, especially, for the latter, when macrofossils and sedimentary structures do not give a clear indication (especially in cores). Figure 3.3 shows the potential of palynofacies analysis in coastal facies.

Table 3.2 Distribution of microfossils in sediments (after Bignot 1985)

+ Abundant − Rare \ Generally absent	Ostracodes	Chitinozoans	Radiolarians	Calpionellids	Conodonts	Foraminifera	Dinocysts and acritarchs[5]	Coccoliths	Spores and pollen[5]	Diatoms
Mudrocks (dark)[1]	+	+	−	−	−	+	+	−	+	−
Limestones[2]	+	−	−	+	+	+	+	+	−	\
Sandstones	−	−	\	\	−	−	−	\	−	\
Coal, lignite[3]	\	\	\	\	\	\	\	\	+	\
Siliceous rocks[4]	−	\	+	\	−	−	−	−	−	+

Notes:
[1] Mudrock includes claystones, mudstones and clayey siltstones ± calcareous varieties
[2] Including chalks. Fossils much reduced in dolomites
[3] Increasing thermal grade leads to decrease in recovery
[4] Of primary nature: chert, flint, diatomite
[5] Organic-walled microfossils are destroyed by weathering, which can penetrate deeply

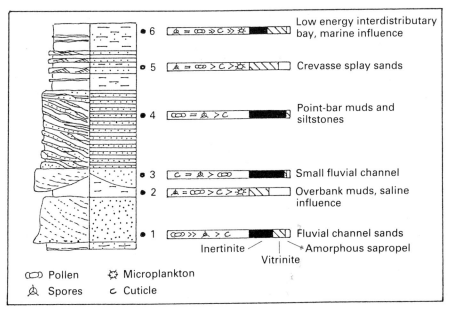

Figure 3.3 Simplified palynofacies analysis of deltaic distributary – interdistributary bay environments. Basal channel sands (1,3) with dominant inertinite (e.g. charcoal). Muddy sediments above (2) contain some marine microplankton. Point bar facies (4) contain inertinite, with cuticle and vitrinite increasing upwards. In (5,6) marine microplankton indicate lagoonal rather than lacustrine environments and closer analysis indicates degree of salinity. On sedimentary structures alone the marine influence would possibly be unrecognised. (Based on Jurassic of Yorkshire, England, Fisher and Hancock, 1985. Note that there are other classifications and that oil yield is proportional to amount of sapropel).

3.5.2 Condensed deposits

Condensed deposits require particularly careful sampling for stratigraphic purposes. With condensed mudrocks, mark and map the rock face. Note carefully from where each sample is extracted, and number each sample immediately. Often there will be subtle changes in lithology, associated with differences in mode of preservation. In condensed limestones, adopt the same procedure and collect large and overlapping samples for later subsampling (*see also* §8.4.2).

3.5.3 Fossil plants

Sampling techniques for *fossil plants* pose more problems than for autochthonous animal remains since, apart from horizons with just roots, autochthonous plants with roots and stems are uncommon. Use can be made of drifted plant accumulations (generally good in type 10 deposits, §1.3) if the limitations of such material are appreciated. Results differ from those obtained by plant ecologists because:

1. The cover does not represent a living community.

2. The cover is unlikely to be a single bedding surface: rocks rarely split perfectly.
3. The quadrat can rarely be placed in a random manner.
4. The way the rock splits is influenced by the type of plant present. The 'cover' for each taxon relates only in part to the original biomass and, in part, to the preservation potential and hydraulic properties of the original plants. If analyses made for sections of similar facies show consistent plant assemblages, then the assemblage is a positive feature of that facies, and can be used in environmental interpretation.

3.6   Field strategy for dominantly autochthonous buildups

Autochthonous buildups include reefs, patch reefs, banks, bioherms, biostromes (Table 3.3). The principal aims are to determine the overall morphology, detailed anatomy and evolutionary history of the buildup.

3.6.1   Potential problems

1. Organisms responsible for buildups have shown a remarkable diversity and undergone much change over geological time (Fig. 3.4),
2. Some organisms (e.g. scleractinian corals) have skeletons composed of relatively unstable mineralogy (aragonite) which readily undergoes dissolution or replacement (Appendix D), and in others skeletal elements are only loosely bound together during life and dissociate quickly on death, and are then easily transported. Bedding and anchoring devices (crinoid roots) may provide clues to the mode of accumulation.
3. There is the problem of unravelling the stratigraphy of a complex buildup and determining time lines. It is important to know what the relief of the 'reef' surface was like. Search for truncation surfaces and local blanketing. Relief may be due to a pre-existing structure, or have been enhanced by differential compaction. Penecontemporaneous erosion may have taken place during formation (check for truncation, evidence of emergence).
4. Any but the smallest buildup displays facies differentiation, e.g. core and flank facies. Within these facies there will be further differentiation, e.g. intraframework sediment, local patches of skeletal sand, local cavities.
5. In well-cemented, and particularly in micritic buildups, it will not be possible to resolve the facies or the composition in detail in the field. Sample for slabbing and preparation of peels and thin sections, using a regular pattern in order to detect facies differentiation.
6. As many animals grow, their skeletons come under attack from a range of organisms, from algae to clams, that penetrate exposed skeleton (e.g. undersurface of corals, or dead areas), thereby tending to destroy the

Table 3.3 A classification of buildups (see also limestone classifications)

A. *Structure*

	Matrix	% Allochthonous material	Diversity	± Talus	% Cavities
Cement reef (skeletons cemented)	low	low	generally low	may be present	generally low
Frame reef (skeletons with local contact)	variable	can be high	generally high	present	can be high
Cluster reef (skeletons generally discrete)	high	can be high	generally high	common	can be high
Stromatolitic/ thrombolitic	micrite	low	low	channelized	variable
Mictritic (mud mound)	—	low	variable	absent	variable
Loose material: crinoidal, fusulines, nummulites,[1] branching rhodoliths	high – low	variable	low	absent	can be high but even size

Note:
[1] mainly allochthonous

B. *Relief and facies differentiation*

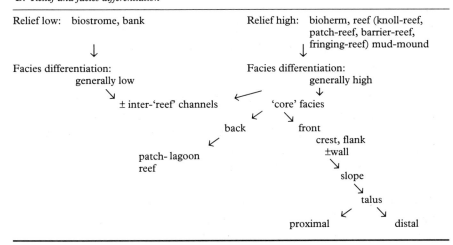

Relief low: biostrome, bank

Relief high: bioherm, reef (knoll-reef, patch-reef, barrier-reef, fringing-reef) mud-mound

Facies differentiation: generally low

Facies differentiation: generally high

± inter-'reef' channels

'core' facies

back   front

patch- lagoon reef

crest, flank ±wall

slope

talus

proximal   distal

buildup. Borers are generally obvious in substantial skeletons, but burrowers can almost completely disrupt more fragile skeletons such as laminar rhodophytes.

7. Skeletal buildups are rather porous. They often formed close to sea level. Fluctuations of sea level may have led to flushing by seawater or freshwater leading to dissolution or dolomitization. Extensive

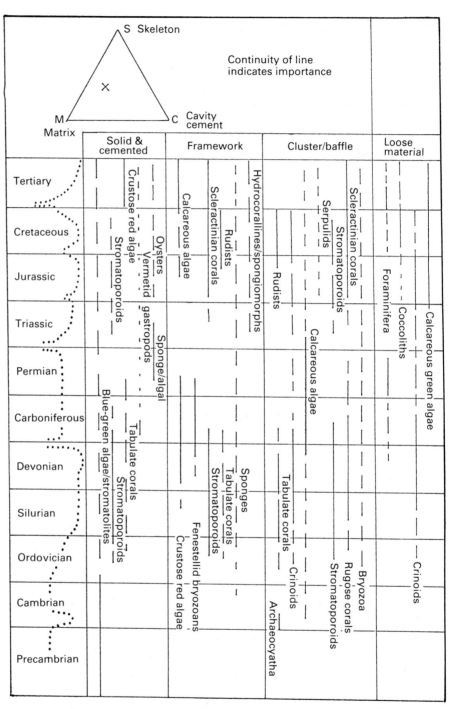

Figure 3.4 (Above) Distribution of organisms through time that form: (1) autochthonous and parautochthonous buildups (an attempt is made to divide the groups on their role as forming (a) solid and cemented buildups, (b) frameworks, (c) clusters and baffles); (2) autochthonous and allochthonous buildups of loose skeletal material (including coccolithic chalk). The dotted

line (left) gives an indication of the importance of autochthonous buildups through time, with periods of their apparent absence. Triangular plot for components of autochthonous buildups (© R Riding, 1987) with x marking plot of buildup depicted on Fig. 1.2. (Below) Generalized cross-section across a carbonate platform and distribution of biotas/facies for the Palaeogene (after Buxton & Pedley, 1989), Triassic (after Aigner, 1985) and Dinantian (Mississippian) (after Simpson, 1987). Modifications to the platform may include: (1) distal steepening; (2) reef belt; (3) outer (rimmed) margin (based on Read, 1985)

dolomitization is a common feature of reef cores and may make work difficult.

8. With larger buildups one is involved with a substantial three dimensional structure. The exposures (quarry, coastal cliff, etc.) may be difficult to work and the degree of weathering insufficient, or joint facies may be spar-covered.

9. Differential growth of the buildup relative to the adjacent areas probably affected turbulence, food availability, light, etc. Anticipate that there may be upward and lateral changes in the composition of the biota, and in growth forms (Figs 6.7, 6.9).

3.6.2 Field strategy

1. Reconnaissance.
2. Map an outline of the buildup at a suitable scale (1:10 or down to 1:1).
3. Map the major facies e.g. core (Fig. 3.5) and flank facies. Tracing bedding surfaces in the flank facies will indicate the topography of the buildup.
4. Establish that there is a major degree of autochthony (§6.2.1).
5. Establish major time lines and event markers, major discontinuities, ash bands, storm layers, emersion events. These provide a basis for unravelling the evolution of the buildup. (NB Autogenic ecological succession cannot generally take place across such events: §6.2.5).
6. Analysis of principal skeletal components:
   (a) Map selected areas using quadrat method (§3.5).
   (b) Attempt to identify the principal skeletal components, their role (Table 6.3) and mineralogy (Appendix D).
   Unfamiliar types of fossil may be present, or unfamiliar modes of preservation. It is important to appreciate the implications of misidentification. For example, a simple alga might have tolerated a wide range of salinities and emergence, but a sponge or coelenterate would have had a much narrower ecological tolerance.
   (c) Identify the growth forms and their ecological significance (Fig. 6.7), determine size and cover of individual components using a gridded overlay. Results may be used to determine percentage vertical and lateral change. Identify the type of substrate below

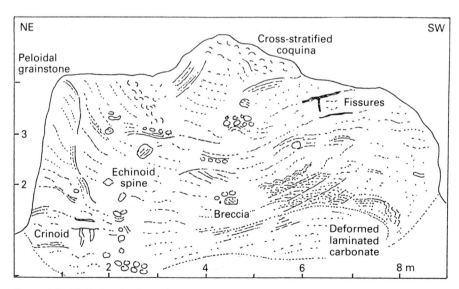

Figure 3.5 Field sketch of core facies of a mud-mound buildup in the Dinantian (Lower Carboniferous) of Derbyshire. It was not possible to establish time lines over the area (after Bridges and Chapman, 1988)

each component, whether formerly soft, or whether one component
is encrusting another, or a clast.

(d) Recognize stages of development (Fig. 6.9), or significant
sequential changes in taxonomic composition and growth form.

(e) Search for growth banding (Fig. 1.2) to establish growth rate and
productivity (which may vary vertically and laterally), and assess
the time represented by the buildup.

(f) Attempt to assess and tabulate the relative resistance to breakdown
of the components (Table 4.2), and the role of skeletal destroyers
(bio-eroders).

7. Interframework biota 'matrix' may be composed of several size fractions,
and include organisms that, mobile or fixed, nestled within the
framework, together with skeletal elements from organisms that readily
dissociated on death (e.g. brittle stars), material washed in (biogenic and
terrigenous), faecal pellets etc. Analysis will allow one to 'complete' the
picture for a reconstruction.

   If the matrix is softer than the framework (as is often the case at
Palaeozoic outcrops), then skeletal components may weather out. There
may be a rich microbiota to sample. If the matrix is harder than the
framework, carefully examine the rock and collect for later slabbing and
sectioning.

8. The nature of cavities is important ecologically and sedimentologically
(Fig. 3.6). Cavities are now either (a) still open/oil filled, (b) spar filled,
(c) sediment filled, (d) crystal silt filled, or these in combination. Try
and determine the three-dimensionality of cavities. (This may require
sampling and slabbing.) First, investigate the nature of any open cavity.
Then tackle filled cavities. Note geopetal structures giving indication of

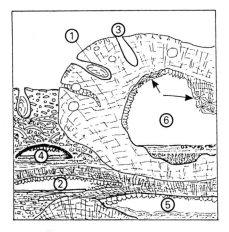

Figure 3.6 Cavities in buildups: (1) primary cavity of bivalve shell and intraskeletal spaces;
(2) irregular growth cavity; (3) borings; (4) shelter structure; (5) compactional cavity;
(6) diagenetic cavity

depositional slope. Primary cavities result from (1) natural cavities (cephalopod camerae, bivalve shell), (2) irregular growth (mm scale for lamellar calcareous algae, to cm–dm scale with laminar corals or stromatoporoids), (3) animal or algal borings, (4) shelter structures. Compactional cavities (5) are common below tabular skeletons and usually display geopetal sediment. Diagenetic cavities (6) occur especially where a skeleton has been wholly or partly dissolved. This may not be easy to prove in the field. Try and determine what was originally present. Look at the margins (arrowed, Fig. 3.6), where impressions of skeletal features may be preserved. A clue to dissolution and replacement having occurred in a rock is an excessive proportion of spar. For instance, common cm-size botryoidal drapes may represent dissolution of algal rhizomes below thin curved laminar crusts. Be prepared for larger scale dissolution features associated with emersion.

9. There is only a limited amount that can be done in the field to appreciate diagenetic history. Try to determine the order in which major dissolution and cementation events occurred. For example, if moulds of framework skeletons are undeformed, then matrix cementation took place prior to skeletal dissolution.

3.7   Bedded fossiliferous sediments

The first objective, following mapping and logging, is to establish whether the units contain autochthonous, parautochthonous or allochthonous fossils (§6.2.1). Most outcrops of lithified sediment are of two types: those displaying bedding surfaces, or those with sections normal to bedding. The information that is potentially available from each is different but complementary. In soft sediments sections are nearly always normal to bedding.

The potential information from bedding surfaces (including quarry floor, shore platform) includes: recognition of colonization surfaces and smothered (obrution) assemblages, analysis of fossil articulation, orientation, pockets, lenses, clusters, lateral variation, diversity, analysis of trails and bedding parallel trace fossils and the opportunity for collecting larger and flatter fossils.

Sections normal to bedding provide information on the detailed stratigraphy, opportunity for spot and channel sampling, examination of tiering of body and trace fossils, shell articulation and orientation and vertical and lateral changes in the biota.

3.7.1   Potential problems

1. In sections normal to bedding, a well-cemented limestone does not display the fauna adequately or its distribution. Sample oriented blocks for slabbing.
2. If only sloping dragline surfaces are available, trench for the stratigraphy

and attempt to split blocks for bedding information.

3. Where bedding surfaces are found only as horizontal clefts then extending the upper and lower faces means, in fact, long-wall mining: a dangerous operation.

4. Bedding surfaces found only as vertical clefts can be extended with care.

5. Where scree has to be traversed in order to examine an exposure, note that material may be falling from above. Check that no danger is being created below when clearing the face. Do not trench in loose sediment without adequate precaution against cave-in: collapse occurs more quickly than one can evacuate.

6. If the 'bedding' surface is a dissolution surface (pseudobedding) then search for true bedding.

3.7.2   Field strategy

1. For allochthonous accumulations (§1.3) collect bulk samples for later analysis of frequency, diversity and 'time-averaging'. For information concerning the source of the fossils, it may be possible to locate a site where the elements are autochthonous. Also search for information that gives a clue about derivation such as attached sediment, type of sediment infilling cavities, and attempt to infer the life environment from morphological evidence.

2. For autochthonous and parautochthonous accumulations (§1.3) seek as much ecological information as possible by systematically checking through (§6.2.3). Record observations on life attitude, and type of trace fossil (§6.2.4). Collect bulk samples for analysis (size distribution, morphological information, growth rates, trophic structure, biomass). Search along section for rarer elements that may be important in upper trophic levels. In facies where shells weather out readily (e.g. muddy limestones) collecting is easy. But consider the validity of any such collections for community analysis. With extensive surfaces quantitative analysis is required for analysis of abundance, dispersion, diversity, population structure.

3. Where bedding surfaces are extensive, lateral variation in the biota is to be expected. At hardgrounds look for crevices and pockets that may yield a different but related biota. Attempt to establish, from trace fossils and encrusters (Fig. 7.17), whether or not a hardground displays evidence for gradual lithification and continuous colonization with gradual faunal (autogenic) succession. Most shallow water hardgrounds display faunal replacement indicating, for instance, an emersion event. Be prepared for larger scale features associated with emersion: caves and cave deposits.

4. Much information can often be obtained from scree material, or boulders on a beach, but these have to be 'fitted' to the stratigraphy.

References

AIGNER T 1985 Storm depositional systems: dynamic stratigraphy in modern and ancient shallow-marine sequences. *Lecture Notes in Earth Sciences* **3**, Springer, 174 pp

BIGNOT G 1985 *Elements of Micropaleontology* Graham & Trotman, 217 pp (With advice on sampling and good review of microfossils)

BOYLES J M, SCOTT A J, RINE J M 1986 A logging form for graphic description of core and outcrop. *Journal of Sedimentary Petrology* **56**: 567–8

BRIDGES P H, CHAPMAN A J 1988 The anatomy of a deep water mud-mound complex to the southwest of the Dinantian platform in Derbyshire, UK. *Sedimentology* **35**: 139–62

BUXTON M W N, PEDLEY H M 1989 A standardized model for Tethyan Tertiary carbonate ramps. *Journal of the Geological Society* **146**: 746–8

FISHER M J, HANCOCK N J 1985 The Scalby Formation (Middle Jurassic Ravenscar Group) of Yorkshire; reassessment of age and depositional environment. *Proceedings of the Yorkshire Geological Society* **45**: 293–8

GRAHAM J 1988 Collection and analysis of field data. *In* TUCKER M E *Techniques in Sedimentology* Blackwell, 394 pp

MELDAHL K H 1987 Sedimentologic and taphonomic implications of biogenic stratification. *Palaios* **2**: 350–8

READ J F 1985 Carbonate platform facies models. *Bulletin of the American Association of Petroleum Geologists* **69**: 1–21

SCHINDEL D E 1980 Microstratigraphic sampling and the limits of paleontological resolution. *Paleobiology* **6**: 408–26

SIMPSON J 1985 Stylolite-controlled layering in an homogeneous limestone: pseudo-bedding produced by burial diagenesis. *Sedimentology* **32**: 495–505

SIMPSON J 1987 Mud-dominated storm deposits from a Carboniferous ramp. *Geological Journal* **22**: 191–205

TREWIN N, WELSH W 1976 Formation and composition of a graded estuarine shell bed. *Palaeogeography, Palaeoclimatology, Palaeoecology* **12**: 219–30

Texts on field mapping and surveying

BARNES J W 1990 *Basic Geological Mapping* 2nd edn Geological Society of London, Halstead Press, 112 pp

BUTLER B C M, BELL J D 1988 *Interpretation of Geological Maps* Longman, 236 pp (Includes summary of the rate of present-day movements of the Earth's surface)

COMPTON R R 1985 *Geology in the Field* Wiley, 398 pp (Much information on mapping techniques, but thin on palaeontology)

LEES A 1989 *Introduction au Levé Géologique* Université de Louvain, Louvain-la-Neuve, Belgium

MOSELEY F 1981 *Methods in Field Geology* Freeman, 211 pp

4

Taphonomy

'It is the understanding and appreciation of taphonomy that
distinguishes a palaeontologist from a biologist'

4.1   Introduction

Any body fossil that might be found has undergone a number of overlapping, but consecutive, processes (taphonomy) since it was a living organism
(Fig. 4.1). To begin with there is death (it is often clear how this took place,
e.g. by burial or predation) and the decomposition of soft tissues, and the

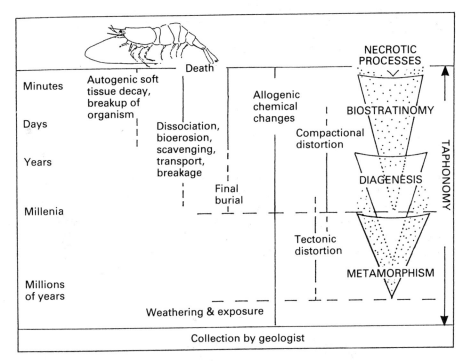

Figure 4.1 Taphonomy and its associated processes

possible break-up of parts. Then follows the sedimentational history, sorting, damage, reworking; stratinomy of the material, leading to its final incorporation into the sediment, and the chemical changes and compaction that take place mainly within the sediment (diagenesis). Tectonic and weathering processes may also occur before the fossil is collected. All these processes tend to modify the original material (Figs 4.2, 4.3).

All organisms are made up of a variety of parts and a variety of materials. The materials (cytoplasm, sheath, aragonitic or calcitic skeleton, chitin, lignin . . .) vary in their ability to withstand degradation, dissolution, breakage or chemical change. The parts (e.g. integument, pygidium, leaf, coxa, dactylus, plate, ossicle) also vary in their ability to withstand dissociation, compaction and entrainment, and transport by hydraulic or wind action.

The eventual mode of preservation also depends on the nature of the sediment in which the organism is entombed: its ability to take impression detail (compare mud to coarse sand), and its permeability and cementation potential. Also important are the preservation potential of the sediment, and the

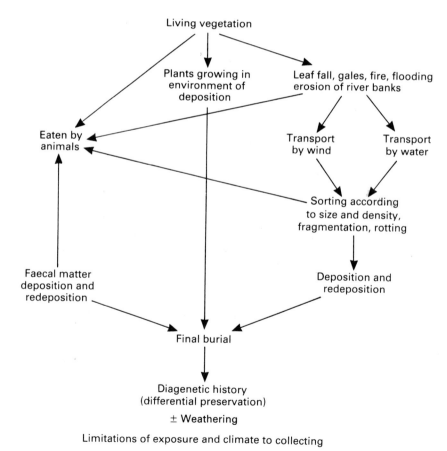

Figure 4.2 Simplified outline of vegetation taphonomy

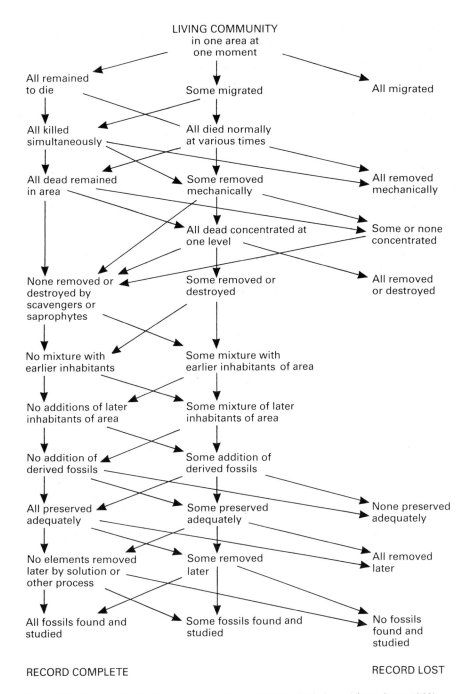

Figure 4.3 Aspects in the taphonomy of a living community (adapted from Ager, 1963)

basinal and tectonic histories of the site, particularly those relating to deformation and to the temperatures that are attained on burial, for example, the thermal alteration of conodonts, graptolites and plant materials (Stach, 1982).

Thus, there are the intrinsic factors relating to the organism, and the extrinsic factors relating to the sedimentary environment and to its geological history. Since sediment accumulating in a specific sedimentary environment is likely to undergo a predictable burial history, any associated biota should have a predictable taphonomy. Fossiliferous sediments that have similar taphonomic histories are referred to as taphofacies (Brett & Baird, 1986). Features to identify include: (a) the likelihood that the sediment may be reworked leading to dissociation and fragmentation, (b) the potential of the sediment to infill cavities, (c) the potential for phosphate and pyrite to form, (d) the compaction potential, and (e) the dissolution and replacement potentials of the minerals involved. The fossils may locally interfere with, and modify, this sequence and gradient because of their large size, porosity (wood, bone), strength and chemistry etc.

Taphonomy can also be used to distinguish between assemblages without necessarily identifying the taxa; a useful tool in facies analysis (Fig. 6.10). This chapter therefore shows how taphonomy can be used in field identification, and in assessing the palaeontological and sedimentological value of any find. Taphonomic processes are often considered to result only in loss of information. This is emphasized by the usual (more than 60 per cent) loss of the soft-bodied (non-preservable) part of the biota. But there are a number of advantageous effects such as enhancement of growth banding in corals, anatomical preservation of plants, enhancement of trace fossils (§4.4), and also time-averaging (§6.2.3.8) and taphonomic feedback (§6.2.3.3).

The mode of preservation should always be considered, in order, for instance, to establish the shape of the aperture of a particular ammonite or, to establish the nature of the stomata of a particular leaf taxon or, to determine the original mineralogy of a shell. On inspection of any fossil, one has to keep asking the question: what additional information is needed and how can it be obtained?

There are four basic modes of preservation of animals and plants: the original material, replacement of the original material, impressions (moulds) of either of these, and encrustation. It is the combinations of these, together with the complexities of the organism and associated skeleton, that so often makes elucidation of preservation mode difficult. But, it is absolutely essential to understand the modes of preservation before any further investigation can be pursued. How the sediment compacted, how the fossil accommodated this compaction (§4.7), and whether or not compaction was arrested by early lithification, as with a concretion (§4.5), are factors that may affect all fossil groups. These are considered below.

Skeletal breakdown on death   In any study of skeletal material consideration should be given to (a) how readily the skeleton may dissociate on death and (b) how readily fragile skeletons (e.g. fragile corals and calcareous algae) may

Table 4.1 A scale of skeletal dissociation

1. Sheathed and spiculate skeletons, where tissues loosely held together e.g. holothurians, some sponges, alcyonarian corals.
2. Segmented skeletons where elements held together by muscle or ligament e.g. crinoids, starfish, brittle stars, echinoids with imbricated plates, echinoid spines, small fish.
3. Bivalved skeletons without interlocking teeth and sockets e.g. razor shells.
4. Bivalved skeletons with interlocking teeth and sockets e.g. cockles, *Trigonia*, terebratulids.
5. Massive skeletons held together by muscles and ligaments e.g. large vertebrates.

be attacked by bioerosion and lose their identity. For each site attempt to determine a scale of dissociation (e.g. Table 4.1). Dissociation is determined by the readiness of the tissues holding elements together to break down, and the way elements may interlock. Bioerosion (*see also* §3.6.2) is a major factor in tropical environments and may radically alter the make-up of the fossil sediment.

Definitions:

Impression includes the surface(s) between fossil and sediment, or between skeleton and skeleton (encrusters), or between skeleton and soft tissue (bio-immuration), as when an oyster encrusts a soft bryozoan.

Replacement includes neomorphism: change of mineral into itself or a polymorph – with gross composition remaining essentially constant.

Natural cast: the form that can be released from a (natural) mould. For example, (1) a mould that results from dissolution of a shell in sandstone may be infilled naturally, (2) decay of the pith may result in the cavity becoming filled to form a cast of the pith cavity.

4.2 Plant fossils

Plant remains vary greatly in their likelihood of entering the geological record. Spores and pollen, produced in vast numbers, are widely dispersed and most are highly resistant to decay, but may be produced seasonally or only once during the plant's life (Palaeozoic lycopods, bamboo). Leaves may be shed or blown from branches still attached to twigs. Flowers may be seasonal and are always delicate. Timber may be transported far out to sea before becoming waterlogged. Plant stratinomy (e.g. Spicer, 1981) is too complex to be considered in detail here, but the plants on a bedding surface are, first of all, the result of stratinomic processes. Fragile material is unlikely to have been transported far.

4.2.1 Vascular plants

Vascular plants exhibit three main modes of preservation. Each reflects the varying preservation-potential (Table 4.2) of the different tissues making up

Table 4.2 Plant fossil preservation modes, and relative resistance of cells to decay

Resistance to decay of plant cells

HIGH	walls of spores, pollen,
↓	cuticle
	lignified (wood) cells
LOW	cellulose e.g. parenchyma

Observations	Mode of preservation	Implications
Absence of carbonaceous, or coaly film	impression	only surface morphology available
Carbonaceous/coaly films	compression, ± impressions	depending on thermal grade, cuticle etc. may be prepared out
(With handlens) cell details evident	anatomical preservation (charcoal, silica, pyrite, carbonate)	cell form and organization available
Associated with concretion	any of above	minimal compactional distortion

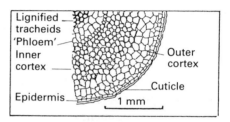

Figure 4.4 The early vascular plant *Rhynia*, to show central lignified tracheids (protoxylem and metaxylem), then 'phloem', surrounded by loosely organized parenchymal cells of inner and outer cortex, and flattened cells of epidermis with cuticle (Rhynie, Scotland, Lower Devonian). Scale bar = 1 mm

the plant. Most of the problems that one experiences when trying to under-stand exactly what is represented by a particular plant fossil, are due to the often complex organization of the plant internally (Fig. 4.4), and between root, rhizome, stem, bud, leaf, flower, cone, seed etc. Effects of compaction, and the way which the sediment splits, have also to be considered. Buried plant material may decay quite slowly so that the processes of decay, sedimen-tation and compaction are more or less concurrent. A Carboniferous log, or piece of peat, will have undergone compaction, probably by a factor of 6, or greater. Logs in a typical Upper Palaeozoic fluvial conglomerate can be compressed to slivers, albeit cm–dm across, originally of similar thickness. But, similar post-Palaeozoic lithologies generally display thicker logs because of the woodier stems prevalent from the Mesozoic onwards.

Impressions (Fig. 4.5)   Quality of the impression depends much on the grain-size of the sediment. Some leaf material may still be present as coaly remains or as carbonaceous upper and lower cuticles (if burial and tempera-ture have been sufficient), but generally will have oxidized away. Sometimes,

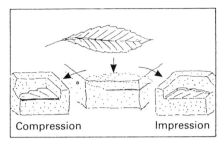

Figure 4.5 Leaf burial leading to compression and impression fossils

a thin bacteria-induced encrustation may have coated the leaf just before burial, enhancing the detail. More complex impressions result from a stem that is decorticated (has lost its bark, Fig. 4.6) or was already partly rotten before burial. Hollow stems, e.g. horsetails, can also become sediment-filled (with sand or compactable mud). With leaves, latex pulls can be taken from the impression of the lower cuticle (bearing stomata) for later examination under electron microscope. Tufa often provides good impressions of leaves, stems and seeds; but irregular calcareous crusts on stems, probably formed in a desiccating environment, are generally unhelpful.

Compressions (Figs 4.5, 4.8) In compressions of leaves all that remains is coaly matter between the superimposed cuticles. Generally, the sediment splits between one cuticle and the sediment impression. It may then be possible, if they are not too carbonized, to lift off the cuticles and prepare them in the laboratory. If the leaf had strongly-curved margins, or was of complex shape, the sediment may not split readily to display a complete surface. Inspect a broken edge to detect the three-dimensional form, and the way the sediment splits. It is important to collect part and counterpart. Compressed megaspores may be readily seen in late Palaeozoic durainous coals.

Figure 4.6 Sandstone cast *Knorria* (Upper Devonian, of decorticated lycopod root North Devon, England). Scale *Stigmaria* with rootlets, bar = 1 cm sometimes referred to as

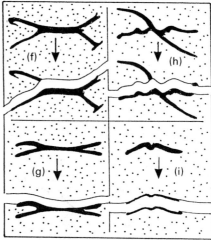

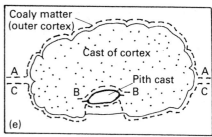

Figure 4.7 (a–d) Preservation of *Stigmaria* (lycopod root): (a) in life with central stele (wood), surrounded by inner cortex (thin-walled cells) and outer cortex (thick-walled cells); (b) death, and decay of inner cortex, collapse of stele to floor, and space filled (c) with cross-laminated sand, before (d) stele and outer cortex compacted to compression state; (e) Stigmarian axis, as in (d) with possible fracture planes, (A–A) passing through outer surface of axis, (B–B) passing over compressed stele and, (C–C) exposing outer surface of axis and then jumping to expose surface of stele compression; (f–h) Lycopod leaves (*Cyperites carinatus*) as compression fossils in mudstone (but still exhibiting some degree of three-dimensionality when viewed in vertical section) with possible fracture planes yielding two-dimensional fossils; (i) Two-dimensional type (*Lepidodendron wortheni*) and fracture plane. Specimen illustrated with incomplete compression i.e. fracture surface has passed through compression (after Rex, 1983, 1985)

Anatomical preservation relates to the three-dimensional preservation of the cellular structure of the plant, by for example charcoal formation or calcification. Decay and compression may have affected the plant before initiation of the processes leading to anatomical preservation. Modes of preservation include:

(i) Charcoal (fusain) formation: this is formed by rapid but incomplete combustion, in partial lack of oxygen, and due to lightning strike, wildfire and volcanic action. It is a strong but brittle material. Twigs, logs and even leaves, flowers and pollen can be charcoalized. Probably most fossil charcoal formed on the plant in its living position. Fusain is a common component of fluvial sediments. The sooty fusain of coal represents pieces of charcoal in the swamp peat.

Figure 4.8 *Alethopteris* as compression fossil in ironstone concretion. Because the concretion formed at an early stage the three-dimensionality is largely preserved (Westphalian). Scale bar = 1 cm

(ii) Stabilization by minerals (e.g. calcium carbonate, silica, pyrite): this occurs in the walls of plant cells (Figs 4.9, 4.10); or crystallizes within the cell itself (permineralization). This is the most valuable mode of preservation for the palaeobotanist. If the cell wall is still present as organic material, and has not been fully replaced (petrifaction), then acetate peels can be made. The dead cells of wood (xylem) lend themselves to conducting fluids, and indeed, in volcanic regions (e.g. Mount St. Helen's, USA) the wood is often partly silicified (like a plumbing system becoming furred) during the life of the plant. Rarely, cells that were once living have also become permineralized with the preservation of cytologic structures. Silicification is also found in sandy facies where silica is the main cementing mineral. But there are exceptions, with silicified plants occurring in carbonate sediments (e.g. Purbeckian of southern England).

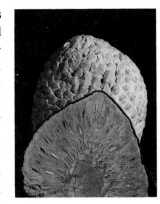

Figure 4.9 Silicified cone of monkey-puzzle tree *Araucaria*. Exterior and in longitudinal section (Jurassic, Patagonia). (× 0.5)

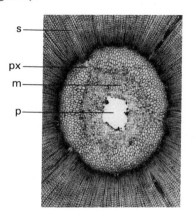

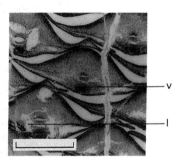

Figure 4.10 *Lepidophlois kansunum*, (left) inner part of stele with pith (p) (largely lost), metaxylem (m), protoxylem (px) and secondary xylem (s) (×5); (right) leaf cushion, ligule pit (1), vascular trace (v) and lateral ventilating tissue (×2), Pennsylvanian, Kansas.

Calcification can be more pervasive. Although best known from the calcified peats (coal balls) of the Silesian (Pennsylvanian), it is widely distributed in associated volcanic facies. Keep a lookout for calcified twigs or roots in non-marine facies. Often silicification has been so rampant as to destroy much of the anatomical detail and what, in the field, might seem promising material may, in thin-section, be disappointing. Be on the lookout for material for growth-ring analysis. Thicker-walled late growth is generally the better preserved.

In sandy or silty sediment, pyrite may nucleate on plant stems and seeds, or within woody cells. A stem may pass through a pyritic concretion. This type of preservation may look poor, but under the scanning electron microscope, details may be present. Similarly, limonite or phosphate may be present.

Impression, compression and charcoalified (fusainized) plant fossils may be associated with concretions (Fig. 4.8). The three preservation modes may intergrade. For instance, charcoal may be incorporated into a coal ball, or a stem in a coal ball may continue beyond the coal ball as a compression fossil.

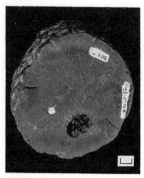

Figure 4.11 *Stigmaria* (lycopod-root). Inner cortex rotted and space became filled with sediment, but xylem rotted later and remained unfilled. Outer cortex largely compressed. Traces of rootlets entering xylem cylinder at rays (Westphalian). Scale bar = 1 cm

Logs of wood may have become partly compressed prior to silicification (or calcification). Wood tends to be compressed parallel to the cell files. Autochthonous roots, upright stems and trunks are generally preserved as impressions of the bark, and compressions of what tissues remained (mainly xylem) after decay and infilling of the space by sand and mud. In a lycopod (Figs 4.7, 4.11), the woody outer cortex becomes compressed, and the small woody stele is often displaced. Particularly difficult to unravel in fossil plants are the different appearances that can occur between different modes of preservation of the same species.

Most aerial parts of fossil plants are found as fragmented and dispersed remains. How can these be reassembled as a reconstruction of the original plant? Palaeobotany is much concerned with this detective work, and it is important to realize, in this respect, the significance of any material one comes across in the field. There are three methods of reconstruction:

1. Locating material where, by good chance, two or more parts (organs) are joined, e.g. root and stem, cone with pollen, and dispersed pollen, leaves and stem, leaves and reproductive organs. This is the only positive method, and it depends mainly on field observation.
2. Identifying particular anatomical features that are common to two or more organs. Frequently-cited characters are particular glands, type of stomata, cell outline: features that cannot readily be identified in the field.
3. Commonly occurring associations of organs (e.g. on bedding surfaces) that would be difficult to explain if not from the same plant.

4.2.2   Coal and oil shales

Coal is described in terms of rank (thermal grade – Fig. 4.12) and type (composition). The end members (anthracite and peat) can be fairly readily identified in the field. Very bright, splintery coal is likely to be of high rank (low volatile bituminous coal-to-anthracite). The tectonic – sedimentary setting of the site, and the grade of the associated mudrocks, will provide a clue to rank. Is there a root bed present? Is the coal autochthonous or drifted? How 'pure' is the coal? It may be no more than a carbonaceous mudstone. Try to identify the components of a bituminous coal:

> *Vitrain*, representing coalified logs or pieces of bark, is glassy (vitreous) and usually closely jointed.
> *Fusain* generally forms lenses and has a silky sheen, but is soft and leaves a black mark.
> *Durain* is dull and tough, with megaspores often evident in Palaeozoic coals.
> *Clarain* (attrital coal) is finely-laminated vitrain and durain.

This classification of coal is tedious to follow for logging. Try using mm, cm

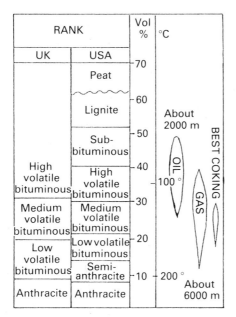

Figure 4.12 Coal rank, volatiles (%), classification and relation to oil and gas generation, and coking potential (adapted from Stach, 1982)

or dm thicknesses of vitrain, fusain or attrital coal. Further detail can be introduced by making estimates of vitrain:attrital ratio. What one is after is an indication of the overall vitrain:fusain ratio, and the percentage of mud present.

Oil shales and cannel coals have formed in aquatic anaerobic environments. Cannel coal (dull with conchoidal fracture) represents drifted, finely-divided terrestrial plant material. It is often found at the top of a coal, following a flooding event. Torbanite, lamosite, tasmanite and marinite are oil shales distinguished by their principal components (identifiable microscopically): torbanite, *Botryococcus* and allied algae; lamosite, planktonic algae (both are lacustrine); tasmanite with *Tasmanites* (marine); and marinite with various planktonic marine algae (Hutton, 1986).

4.2.3   Calcareous algae and stromatolites

The undulose, arched or planar laminations of stromatolites (Fig. 4.13) are generally readily recognized in the field. The laminations reflect discontinuities in formation (e.g. storm action) rather than short (e.g. diurnal) periodicities. Nodular stromatolites (oncolites) are distinguished by their irregular lamination around a shell or lithic clast. The clotted texture of thrombolites is not easy to prove in the field.

Growth form is probably the best indicator of red, calcified algae with forms corresponding to the shapes of most types of breakfast cereal. It may just be possible to make out, with a hand lens, the cellular structure of a solenopore, or the larger, reproductive cells of more advanced rhodophytes. Rhodophytes

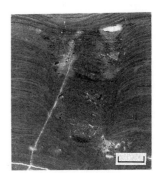

Figure 4.13 *Kussiella kussiensis* (stromatolite) (Early Riphean, S. Urals). Scale bar = 1 cm

will probably have an irregular, knobbly outline attributable to truncation and branching.

Although the stem of the green alga *Chara sensu lato* may become encrusted, it is the calcified walls of the cells enclosing the oogonia, the gyrogenites, (often replaced by silica, or seen as moulds) that are so useful as environmental indicators (essentially fresh water). Use a hand lens to identify the spiral cells. Marine, calcareous (aragonite) green algae such as *Halimeda* readily breaks up into segments, as does *Corallina* (articulated, red alga) (*see also* §3.6.1).

4.3   Animal fossils

4.3.1   Soft tissue preservation

Soft tissue preservation is rare (*see also* §6.3). In the field consider what types of sediment may yield information on such material. There are five modes of preservation: (1) mineral coats, where tissues are outlined by a mineral, typically pyrite; (2) permineralization, actual replacement of the tissue by phosphate (best), pyrite or silica (as with plant cells, above); (3) casts or impressions of the tissues, where the adjacent sediment was stabilized prior to decay and infilling of the void: or, for example, by concretion formation in the Silesian Mazon Creek Formation, but probably by organic stabilization in the late Precambrian Pound Quartzite (Fig. 4.14); (4) encasement in decay-inhibiting material; for example, scorpions in Carboniferous coal (peat), insects in amber; and (5) encrustation of soft tissue (e.g. hydroids, algae) by skeletal organisms, to leave a mould of the soft tissue (bio-immuration).

Evidence of bioturbation, scavenging and early reworking almost always excludes soft tissue preservation of types 1–4 (except where concretion initiation was so early as to inhibit these factors locally). Although transport *per se* prior to deposition does not appear to be a major factor, aerobic decay during transport can be important. Following deposition, it is early diagenesis that is important since anaerobic decay may be complete (Fig. 4.15).

Figure 4.14 *Dickinsonia* (Ediacarian, South Australia) on sandstone sole, as concave hyporelief. Scale bar = 1 cm

Note that:

1. In freshwater sediments, pyritization of soft tissues is relatively unimportant due to relatively low levels of sulphate in the water. In marine sediments with sufficient iron, pyritization is favoured by high burial rate. With rapid burial more reactive (less decayed) organic matter is added to the sediment, leading to a rapid uptake of sulphate ions from pore water and therefore increased diffusion of sulphate ions from the overlying sea water. In euxinic conditions organic carbon is generally too high to favour good pyritization and pyrite is disseminated.

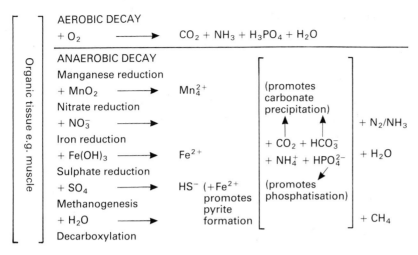

Figure 4.15   Ideal stratification of bacterial reduction zones. Nitrate reduction and methanogenesis dominate in freshwater environments, sulphate reduction and methanogenesis dominate in marine environments. Note that iron (Fe^{2+}) may be available during whole period (adapted from several sources including Allison, 1988, Eglington *et al.,* 1985)

2. Phosphatization of soft tissue is generally associated with low burial rate (but without reworking) and high organic input. Phosphatic materials (bone, arthropod carapace) often act as nuclei, but the best phosphatization seems to have taken place, or at least to have been initiated, at the sediment – water interface (Martill, 1988; *see also* §4.5).
3. Early carbonate encasement, where calcium carbonate was precipitated as concretions, is favoured by high organic input and burial rate. Where terrigenous input is low (marine or freshwater), rapid carbonate precipitation is favoured by high carbon dioxide production by algal and cyanobacterial mats e.g. the bedded limestones of Solnhofen (Upper Jurassic, Germany) and the Green River Formation (Eocene, USA) where fossils always seem to be compressed.

4.3.2   Univalves and groups with skeletons that remain more or less intact

Graptolites (V – revised) As with ammonoids (below), the mode of preservation of graptolites in a sequence of mudrocks may change appreciably from bed to bed, indicating small variations in facies. Graptolites had a pliable organic (collagenous) skeleton. Three-dimensional preservation in early-formed concretions and certain limestones is uncommon. Material can be collected for later acid-treatment. The most common mode of preservation is as carbonized side-down (profile) compressions, or compressions where the organic skeleton has been replaced by white-weathering, phosphatic material (Fig. 4.16). Diplograptids (Fig. 4.17), however, commonly landed on the

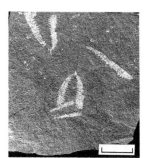

Figure 4.16 *Didymograptus* compressed as white film (Lower Ordovician). Scale bar = 1 cm

Figure 4.17 Diplograptids, with micritic infill, in micritic pebble in conglomerate. One specimen in scalariform view. Scale bar = 1 cm

Figure 4.18 Archaeocyathid. Skeleton slightly etched from matrix (Ajax Mine, Flinders Ranges, South Australia, Lower Cambrian). Scale bar = 1 cm

Table 4.3 Univalves and groups with skeletons that remain more or less intact

Organic skeleton	graptolites
	conularids
Siliceous skeleton	some lithistid sponges
Cephalopods	conch is aragonitic with subsidiary parts calcitic, organic or phosphatic
Calcitic and/or aragonitic	archaeocyathids
	rugose, tabulate and scleractinian corals
	serpulids
	most bryozoa
	sponges
	larger foraminifera
	gastropods
	scaphopods
	rostroconchs
	stromatoporoids

(letters refer to volumes of the *Treatise on Invertebrate Paleontology*, Appendix B)

sediment with one series of thecae being entombed aperture down (scalariform view). If the form of the thecae is clear, identification is generally possible. Locate material with discrete rather than crowded specimens and with the proximal morphology distinct. Take care not to misinterpret apparent branchings for *Cyrtograptus* or *Nemagraptus*. Commonly, pyrite filled the thecae at an early stage. Only internal moulds may remain but these have the advantage that they are undeformed.

Where mudrocks contain distal turbidites (striped mudstones), the graptolites therein tend to be richer and better preserved than in more oxidized sediment. Preservation was enhanced by rapid sedimentation.

Conularids (F) are generally highly compressed in mudrocks. External and internal impressions occur in sandstones.

Archaeocyathids (E – revised; Fig. 4.18) Skeleton generally with morphology distinct. The calcareous skeleton is now usually of calcite or has been silicified, but may have originally been Mg calcite. Geopetal sediment and sparry fill to chambers is common. Check for epitaxial algae (*Epiphyton*) as dark micritic clots on autochthonous material. Mouldic preservation should show evidence of regularly porous walls.

Sponges (E) It is exceptional for a fossil sponge to be found relatively perfect. The best preserved are those with fused mineralized spicules giving a rigid skeleton. (Peripheral, loosely bound spicules will still be missing.) Judging by the abundance of loose spicules in many sediments (e.g. the needle-like spicules in many Jurassic and Cretaceous rocks, or the ball-bearing-like dermal spicules (*Rhaxella*) in the Upper Jurassic) sponges were much more widespread than is now apparent. Opaline spicules of demosponges and hexactinellids are readily dissolved. Look out for moulds (in sand and mudrocks), or carbonate replacement in limestones. Sclerosponges have a basal calcareous skeleton with embedded siliceous spicules. In fossil represen-

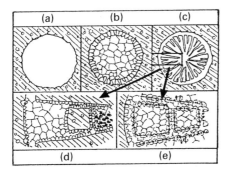

Figure 4.19 Modes of preservation of a scleractinian coral (aragonite): (a) void with impression of external morphology; (b) calcite-filled void: 'cast'; (c) retention of septa by either (d) spar replacement between matrix-filled interseptal spaces or; (e) neomorphic replacement, calcitization (with distinct micritized wall to septa) in a phreatic environment. In (d) a shelter cavity was partly infilled by pellets. It may then be inferred that the wackestone in which the coral is embedded was originally pelletal

tatives (e.g. *Chaetetes*) the skeleton may now be entirely calcitic.

Rugose and tabulate corals (F – revised) In sandstones, preservation is nearly always poor. If the skeleton is present then it will be almost certainly replaced. Sediment may have infilled the calyx and infiltrated more deeply, giving variable perfection to moulds. Preservation in mudrocks can be good especially if they are somewhat bituminous. In wackestones and lime mudstones, compaction is common. Select material suitable for transverse and longitudinal sectioning. The manner in which colonial forms increase is important.

Scleractinian corals (F – Fig. 4.19) Due to the originally aragonitic skeleton, mouldic or replacement modes of preservation are widespread. Neomorphic preservation (calcitization) is mostly associated with massive colonies (because they were less permeable). Colonial and solitary forms can be completely dissolved and the resulting cavity spar-filled. Preservation is generally poor in reefs, becoming better laterally in muddier facies. For full identification it is essential to have good preservation of the skeleton for transverse and longitudinal sectioning. Distinct banding (probably emphasizing original seasonal growth variation) is often evident.

Bryozoans (G and G – revised) The stony bryozoans of the Palaeozoic had a calcitic skeleton and are like miniature tabulate corals, showing similar preservation modes. Preservation is often clear in mudstones with specimens weathering out. Good preservation of external morphology and, then, sectioning is necessary for close identification. Fenestrate bryozoans are typically exposed with the apertures (smaller than the fenestrules) facing down (Fig. 4.20), so it is necessary to search for material with apertures facing up. Slender, encrusting cyclostomes are·easily damaged. In many cheilostomes the frontal surfaces include delicate sculpture, ovicells and avicularia, costae and spines. Some loss of detail may have taken place because of abrasion or dissolution. Also, replacement usually leads to loss of detail, though silicified

Figure 4.20 Fenestellid bryozoan, mainly displaying non-apertural side, but small area right (by sketch) with apertures (Dinantian (Lower Carboniferous), North Wales). Scale bar = 1 cm

bryozoans within hollow flints may be well preserved externally. Impressions of the external surface of Palaeozoic forms are of little significance.

Larger foraminifera (C) Larger foraminifera are common at certain times and in certain facies and are generally well preserved in fresh rock. Fusulines (Upper Palaeozoic), alveolinids (Mesozoic onwards), and rotalids (e.g. *Nummulites*, orbitoids of the Tertiary) all had calcitic skeletons. In limestones, preservation is generally good. Individuals may weather out, or blocks may be collected for later slabbing. In sand and mud rocks, weathering may lead to dissolution. Break some specimens open to check that skeletal detail is present.

Stromatoporoids (F) are a heterogeneous group. Break off a piece of rock and examine the fresh, wetted surface with a hand lens to spot the skeletal laminae and pillars (never present in stromatolites). Thin-sections are required for identification.

Gastropods (I) Most modern gastropods have an entirely aragonitic skeleton. The operculum, when present, is horny or aragonitic. Preservation mode is similar to ammonites but, being without septa, sediment readily enters the spiral shell. In bioclastic limestones and sandstones, and in oolites, preservation is generally as impressions of exteriors, infills (cores) or with replaced shell. With impressions, it is more difficult to locate specimens with the earliest growth stages preserved and the terminal aperture intact. In mudrocks, compressed material can be difficult to determine. Spinose forms will need care in collecting.

Serpulid tubes are common, free or attached, and generally well preserved with the original skeleton, or as mouldic preservation. Many are encrusters but the host may often not be evident if it was aragonitic and has been dissolved. Massed tubes of serpulids can form substantial aggregations. Other annelids line or construct their tubes with a variety of materials.

Scaphopods have always had an aragonitic skeleton (so far as is understood). They are uncommon as fossils until the Tertiary and then can often be found with original aragonite in mudrocks.

Externally-shelled cephalopods (ammonoids – L; nautiloids – K); Fig. 4.21, *see also* §6.1. Nautiloids, ammonites, goniatites and ceratites are among the most common of macrofossils, readily recognized and often exquisitely preserved. Evidence of death is occasionally evident as bite marks. Today *Nautilus* conches are dispersed by wave current action far beyond their 'home waters'. The same is likely to have occurred with ancient cephalopods. Their taphonomy is astonishingly varied and complex. In particular, their chambered organization and, in many instances, their shape, means that local micro-environments, formed during diagenesis, are 'out of step' with the general diagenetic gradient (Fig. 4.15), e.g. pyrite forming in chambers. The different modes of preservation have important implications for palaeontology, stratigraphy and sedimentology.

Whilst there are many variables, identification can frequently be achieved

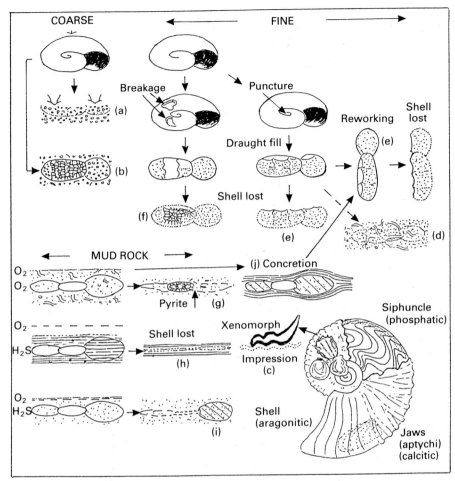

Figure 4.21 General model of cephalopod preservation (For explanation see Table 4.4)

on a few features:

> Palaeozoic nautiloids: form of conch, nature of siphuncle and cameral deposits, form of aperture.
> Palaeozoic ammonoids: suture line, perforate or imperforate umbilicus, form of conch and venter, lateral sculpture.
> Mesozoic ammonoids: form of conch and venter, ribbing and tuberculation, suture line.

For information on ontogeny, pyritized or phosphatized inner whorls are needed. Where reworked cephalopods are present, take care in making stratigraphic conclusions. Take advantage of any differences in preservation mode that may be present, which can provide information about the early diagenetic regime.

The factors that seem to have most influenced the mode of preservation are shown in the following flow chart:

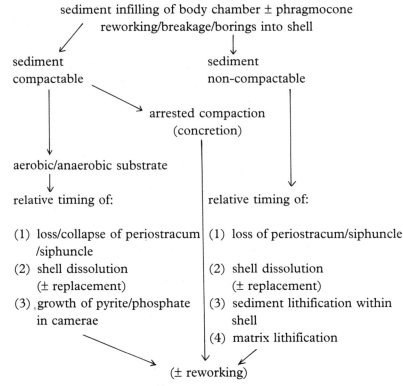

It is generally possible to identify by which of the four methods the shell was infilled by sediment:

1. Direct filling of body chamber.
2. Damage to, or boring through outer wall, leading to filling of connected chambers.
3. By draught fill through septal necks (following damage to, decay of siphuncle, thus allowing a current to flow through the shell and deposit sediment).
4. Sediment entering, following partial dissolution of shell.

Other aspects that have influenced mode of preservation are:

1. Preburial encrustation by oysters, serpulids (*see* xenomorphic preservation below), stromatolites.
2. Attitude of shell on embedding, i.e. vertical or horizontal.
3. Shell form (involute: inner chambers protected; evolute: inner chambers unprotected; strong keel, e.g. *Asteroceras*; rounded venter, e.g. *Nautilus* leads to shell fracturing irregularly, and to lateral spread under compaction).
4. Pyritic film on outside (e.g. giving limonitic film on many Chalk ammonites).

5. Growth of carbonate or pyrite in chambers or infilling of chambers by hydrocarbons (Fig. 4.22).

6. The thinner shell of the first formed whorls which was more readily damaged.

In the Palaeozoic, many orthocones had a complex siphuncle and careful collecting is required to determine this. The great length of some means that a specimen might not be completely enclosed by a concretion. Note that pressure dissolution (stylolite) often leads to loss of information on suture line and sculpture.

The sedimentological and palaeontological implications of various preservation modes are shown in Table 4.4. There are four main preservational regimes (based, in part, on Seilacher, 1971; 1976):

Figure 4.22 *Harpoceras* with some camerae filled with bitumen, within a concretion (Lower Jurassic). Scale bar = 1 cm

1. In sandstones, ammonoids are generally found in mouldic preservation.

2. In oolites and bioclastic limestones (a,b), compaction is not an important factor. But what is important is whether shell dissolution preceded (a) or succeeded (b) lithification. In the former, sediment more or less just dropped into the camerae as septae and outer shell dissolved. Sometimes, all that remains is an impression of the trough-form of the underside of the shell. (Such 'half-ammonites' can also be formed by later dissolution.) But, if lithification was early, only the body chamber was filled with loose sediment, and the camerae (unless damaged) remained available for possible sparry fill. The shell was then replaced.

3. Micritic limestones (c–f) show a puzzling range of preservation mode. In Cretaceous chalk, since the substrate was often below the aragonite dissolution depth, ammonites (and gastropods and corals) are preserved only at specific horizons. Concentrations of original calcitic aptychi in Tethyan pelagic limestone are due to dissolution of the conch on the substrate. If the shell was draught-filled and buried, and dissolution took place but without lithification, then burrowers were able to pass through. The fill was subsequently deformed by compaction, and septal traces closed (and were lost). A composite mould may also result (d). In many sequences of condensed limestone (such as the Tethyan Ammonitico Rosso) and hardgrounds, reworking or, at least, re-exposure was common. If the draught fill was lithified prior to reworking, as in the Triassic Muschelkalk, the cephalopod was no more than a special type (e) of intraformational clast (Fig. 4.23). In the micritic limestones of Solnhofen (Upper Jurassic), ammonite preservation resembles that of bituminous shales: compaction took place as the shell dissolved. (Note: Draught filling of the phragmacone is usual, but not always possible to prove because of later compaction.) For phosphatic and glauconitic concretions, see §4.5.

Figure 4.23 Ceratitic ammonoid with draught fill and geopetal structure. Shell subsequently dissolved (Middle Triassic, Germany). Scale bar = 1 cm

4. In mudrocks (g, h, i) lithification was delayed so that compactional effects, growth of pyrite and phosphate, and the formation of

Table 4.4 Observations and implications of ammonoid taphonomy

Observations	Sedimentalogical implications	Palaentological implications
Oolites and bioclastic limestones (a) Shell lost, ± impression of lower surface only. Large biocasts in infill	Early dissolution of shell	Loss of information of external sculpture and suture lines; loss of ontogenetic information; loss of body chamber and apertural form; loss of cross-sectional form
(b) Conch preserved (replaced); conch uncompressed; camerae spar-filled and/or draught-filled or, with damage to phragmacone, matrix-filled	Matrix lithified before shell dissolution	Good preservation of sculpture, growth lines and of suture lines (when shell removed)
Micritic limestones (c) Attachment impression, xenomorphic features and/or aptychi (no aragonitic fossils)	Substrate below aragonite compensation depth	Loss of palaeontological and stratigraphical information
(d) Bioturbated (and often deformed) internal/composite mould. Sutures ± lost	Shell dissolution and plastic deformation. Delayed lithification	Only gross features remain
(e) Uncompacted body chamber and phragmacone as internal mould	Early lithification of fill	Loss of information on sculpture, and generally some loss of suture lines
(f) Micritic fill to some chambers, others missing or spar-filled (i.e. local fill following damage to shell)	Early lithification of fill, ± spar formation	Preservation variable, search for suitable material
Solnhofen mode (as for bituminous shales)		
Mudrocks (g) Bioturbated mudrock Pyrite associated with ammonoid; early camerae with pyrite infill	Pyrite/organic content low, but local environments of sulphate reduction in camerae; aerobic substrate	Good preservation of inner whorls and possibility of ontogenetic studies. Often difficult to locate outer whorls
(h) Bituminous shales High organic content; ammonoid completely flattened; aragonite dissolved, ± thin periostracum film; pyrite disseminated in sediment	High compaction; anaerobic substrate	Identification difficult if crushed. Sculpture distinct only when early fracture of body chamber has occurred. Phragmacone dissolved rather than crushed. Loss of suture line usual
(i) Restricted mudrocks Pyrite low to absent; aragonite shell generally present (or replaced), but crushed beer-mat preservation; ± concretion (phosphatic or calcareous), body chamber generally evident (Fig. 4.24)	Aerobic layer just below substrate; intermediate organic content	Often possibilities for ontogenetic studies when inner whorls pyrite lined (not sediment filled). NB: Body chamber concretion may not extend to aperture

concretions were the important factors. Pyrite that nucleated on the cameral walls hindered compaction, allowing, especially, preservation of inner whorls (g). While body chambers readily became filled by sediment, unless a concretion formed (§4.5), compaction took place (Fig. 4.24). Generally, the environment was too quiet for draught filling to occur. Cephalopods that came to rest on muddy substrates were natural 'islands' for settlement of hard-substrate oysters and serpulids. Since these are calcitic, attachment areas and xenomorphs (c) may be the only evidence. Note that most oysters with impressions or xenomorphs of ammonites in mudrocks result from aragonite dissolution in the weathering zone.

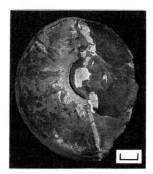

Figure 4.24 *Harpoceras* in body-chamber preservation, with rest of shell flattened (Lower Jurassic). Scale bar = 1 cm

Belemnites and other coleoids Every geologist is familiar with belemnite guards, and the calcitic guard or rostrum with part of the aragonitic chambered phragmacone is also common. But the anterior extension of the phragmacone (pro-ostracum), which was not always completely mineralized, is uncommon. The phragmacone and pro-ostracum are nearly always crushed where unprotected by the rostrum. To find these with the ink sac and tentacles (with chitinous hooks) all together, is so rare that composites have been made. The chitinous arm hooks may be spotted in clayey sediment.

4.3.3 Groups with skeletons that separate into two parts

This group largely concerns the bivalved shells: brachiopods, Bivalvia, ostracodes, branchiopods (Fig. 4.25).

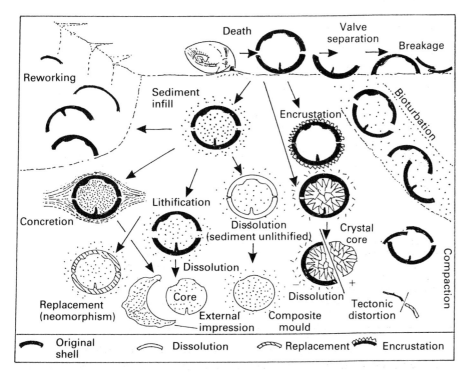

Figure 4.25 General model for the taphonomy of a bivalved shell

Figure 4.26 Decalcified fine sandstone with moulds of orthid brachiopods and tentaculitids, together with plastic replica (Ordovician). Scale bar = 1 cm

Figure 4.27 Partly decalcified calcareous sandstone with dissociated brachiopods: 'blue-heart' preservation (Ordovician). Scale bar = 1 cm

Figure 4.28 Terebratulid brachiopod encased within and without by flint. Shell dissolved (Upper Cretaceous, England). Scale bar = 1 cm

Brachiopods (H) in Palaeozoic sandstone–mudstone–shale facies, representing mostly shallow marine environments, are commonly found as impressions (moulds) of dissociated shells (Figs 4.26, 4.27). A minimum of four impressions is required – of the outside and inside of each valve. Details around the pedicle area and dentition are particularly important. It may be necessary to collect several specimens to make up, say, a complete impression of a brachial valve. Fine- to very fine-grained sandstones provide the best moulds. Avoid wet, weathered or argillaceous sandstones. It is generally possible to get an appreciation of a particular bed by inspection of the weathered edge. Weight of the blocks is also indicative. But, some blocks that look promising from the outside will turn out to be blue-hearted (Fig. 4.27). Non-decalcified material can be collected for later acid decalcification.

In limestones, problems encountered are the welding of the shells to the matrix and the way pseudopunctate shells tend to split within the thickness of the shell (then displaying the pseudopunctae). Details of the umbonal area may be difficult to determine, and the rock seldom breaks to reveal internal structures completely. With joined valves, serial sectioning may be needed later. Many brachiopods have delicate spines and margins. Ensure that all such detail has been collected.

Mesozoic and Caenozoic brachiopods better resisted disarticulation, typically retain their original skeleton and are easily released from the sediment. Inarticulate brachiopods with organic shells are often crushed and crumpled.

Bivalvia (N) There is much variation in the ligamentation of bivalve molluscs and this is reflected in the readiness of the valves to dissociate. Bivalve molluscs show more complex diagenesis than brachiopods. Calcitic oysters are nearly always preserved with original mineralogy. In other bivalves it is the aragonite that is so prone to dissolution. For instance, in Chalk (Upper Cretaceous) *Inoceramus*, the nacreous inner layer, though relatively thicker, is often lost and all that remains is the thinner, outer (and often fractured) calcitic (fibrous) layer (Fig. 4.29). Compaction will have generally closed the chalk about the prismatic layer so that all trace of internal marking and muscle impressions is lost. But specimens can be found where the aragonitic layer has been replaced by calcite. In calcareous sediments, preservation can be better in allochthonous shell beds (which were cemented before aragonite dissolution) than in associated micrites (where the shells were leached prior to cementation). For full identification, complete valves (or equivalent impressions) with dentition, muscle and ligament areas are necessary. Thin-shelled bivalves are liable to breakage/compaction.

Ostracodes and branchiopods (R) exhibit virtually the same preservation modes as brachiopods. It will generally be possible to release the valves from mudrocks and calcareous sediment. Mouldic preservation can be collected for later (latex) casting.

4.3.4   Groups with skeletons that separate into many parts

This group of fossils includes echinoderms, many arthropods, vertebrates, (for sponges see §4.3.2)

Echinodermata (S, T, U) (Fig. 4.30) have only one chance to be preserved complete and that means, in effect, burial in life position or very soon after death. Following death, no reworking (hydraulic or bioturbation) is possible without dissociation taking place. Even without reworking, there is likely to be loss of the smaller skeletal elements such as pedicellaria, the smaller miliary spines of echinoids and pinnules of crinoids. Five grades of preservation of echinoderms can be distinguished (adapted from Smith, 1984):

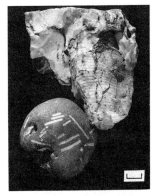

Figure 4.29 *Inoceramus* in flint. In rounded specimen are fragments of the outer prismatic layer. In the other specimen the prismatic layer has been lost revealing the surface between the prismatic layer and the inner nacreous layer, which was lost before formation of the flint (Upper Chalk). Scale bar = 1 cm

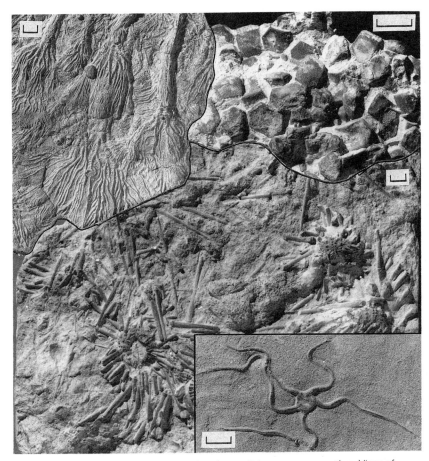

Figure 4.30 Group of *Hemicidaris* in virtual autochthonous preservation. View of undersurface with oral-side-down test on shelly packstone. The group was overwhelmed by oolite (Bradford-on-Avon, England, Middle Jurassic). Scale bar = 1 cm. (Upper left) *Pentacrinus* drawn out, possibly below a sunken log to which they had been attached when log was floating (Lyme Regis, England, Lower Jurassic). (Upper right) Dissociated plates of *Palaechinus* (Dinantian) (northern England). (Lower right) *Palaeocoma* (brittle star) on sole of storm-event bed in autochthonous preservation (Dorset, England, Middle Lias)

1. Near-to-perfect preservation with, for instance, echinoid with spines, pedicellaria, lantern and apical disc intact. This does occur but has required exceptional conditions of smothering (storm, slump). No reworking will have taken place. If compaction has occurred (likely with the empty test), the oral and aboral discs of echinoids will have collapsed first. Crinoid cups were generally stronger.
2. Dissociated skeletons, where plates of the calyx are separated, but the stem and arms belonging to the particular calyx can be recognized adjacent to it. In echinoids, the spines may be seen detached around the test. The stratinomical processes took place under quiet conditions.
3. Good, but incomplete, preservation with loss of arms, and probably, stem of crinoids, and loss of spines, apical disc and lantern in echinoids. This is the most common preservation mode of crinoids, echinoids and blastoids in collections.
4. Preservation as abundant, dissociated, skeletal elements, except that, for instance, a few crinoid ossicles may still be joined. It will be extremely difficult or impossible to reconstruct a whole individual. Nevertheless, identification of class or even genus may be possible.
5. Small crinoids, brittle stars and holothurids readily break up. Their remains may be recognizable only on later microscopic examination of a washed sample.

In the Palaeozoic, many echinoids had an imbricate-plated test; much less likely to be preserved intact than those with rigid tests. In general, epifaunal echinoderms in shallow-water environments have a lower preservation potential than groups that live infaunally or in quieter and aggrading environments. An exception may be found in the sand dollars with their strong, pillared test and deeply inter-penetrating pegs and rods.

Most elements of an echinoderm skeleton are a single crystal of porous stereom. Syntaxial overgrowth is likely to occur in any sediment that is reasonably permeable leading to a readiness to break with calcite cleavage. (Preburial breakage is always irregular across the stereom). In muddy sediment, the stereom may have been penetrated or remain virtually unfilled, though a thin film of iron sulphide is often present. Washings from Tertiary mudrocks yield spatangid spines looking as clean as any modern spine. Palaeozoic crinoid ossicles have been treated in the laboratory with hydrofluoric acid which removed siliceous matter and replaced the stereom with fluorite. In sandy sediment, dissolution leads to moulds. Moulds are common in flint and chert.

Trilobita (O; Fig. 4.31)   Preservation modes are similar to those of brachiopods; the difference being the readiness with which the skeletal elements dissociated. Like brachiopods, it is generally difficult to obtain material from unweathered sandstone or siltstone. Complete specimens are not uncommon, preserved with the original calcite or as moulds. With dissociated material, try

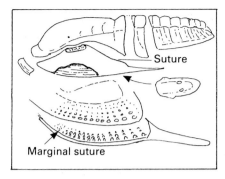

Suture

Marginal suture

Figure 4.31 Exploded trilobites to illustrate potential dissociation of exoskeleton. The trinucleid (below) has a marginal suture.

to obtain the components to make up an 'individual' (though it may not be possible to determine the number of thoracic segments). If several moult stages are present, it is possible to evaluate ontogenetic stages.

Problems arise in collecting mouldic material; for instance, the components necessary to reconstruct the brim of a trinucleid. External impressions of both upper and lower parts (lower part bears the genal spines) are required. Note that the eyes of most lower and middle Cambrian trilobites are ringed by a suture and are generally separated from the free cheek, in contrast to stratigraphically younger trilobites. In fresh limestone, pygidia often seem to be more common than cranidia, but the rock tends to part less readily along cranidia. Compaction in mudrocks can pose problems.

Insects are not so uncommon. If one considers how closely any non-marine environment may be correlated with its insects, then a few insects should be very useful in the interpretation of ancient environments. Insects in amber (the exudative of gymnospermous wood, and equivalent to modern copal) are the best known. Amber is found generally as reworked pieces in terrigenous and shoreline sediments. Keep an eye open for amber in less likely facies e.g. Tertiary flysch sediments, Cretaceous fluvial sediments. Older ambers are important because they can provide information on the evolution of the class alongside angiosperm evolution and diversification. In lacustrine and coal-bearing facies, insects occur carbonized or as impressions (often fragmented) in siltstones. Beetle elytra and cockroach wings are the commonest elements. Preservation of wings differs from leaves in that there is no compression of the membranes (since they are in contact), except along the veins (which may be pyritized). Silicified or calcified material is also known, as well as insect mines in leaves and wood, insect faeces, caddis fly cases, and pyritized pupae. Identification can be difficult: seek expert help.

Crabs, shrimps (R), horseshoe crabs, eurypterids (P) 'Shrimp Bed' or 'Lobster Bed' are common stratigraphic designations and nearly all are associated with fine-grained lithologies. Enclosure within nodules (commonly phosphatic) has enhanced preservation, and distinguishing between corpse and moult can be difficult: the former will have been attractive to scavengers. Fragments are also quite common in a variety of preservation modes, but are often indeterminable even to generic level, because of incompleteness. Nevertheless, even the impression of an elongate chela of a predatory decapod provides a useful bit of ecological information. Chelae are often strongly calcified, thereby raising their preservation potential. Small crustaceans in fine-grained limestones are generally compressed and their cuticle carbonized or replaced, e.g. by fluorapatite (white), silica or calcite. The cuticle may also be invested with pyrite which may intrude the thickness of the cuticle.

Vertebrates   Few of us are going to stumble on, or indeed locate, a large fossil vertebrate. If we do, we should get in touch with a local museum or vertebrate palaeontologist who has experience in extracting such material. Large bones weathering out of softer sediment often provide the first indication. When more complete skeletons are found it is essential that the blocks are extracted with known relationships. More common are dissociated bones and fragments, loose scales and denticles, teeth and otoliths, more or less evenly distributed through the sediment, or hydraulically concentrated: bone beds and fissure fills. This material may be largely first cycle, but also expect reworked material. But, bones and other vertebrate material may be rather dispersed, for example in a conglomerate in which the main components are large logs. In this case consider concentrating the bones by sieving. Reworked material will probably have been altered somewhat especially by the infilling of pores. This leads to sharp, clean fractures in contrast to the irregular fracture of 'fresh' bone (autochthonous or parautochthonous). As with echinoderms, preservation of complete skeletons depends much on facies. Size is also a factor. The upper and lower sides of an ichthyosaur carcass on mud may well have been decomposing under quite different chemical regimes; the upper side aerobic and the lower anaerobic. Professional collectors will always prepare from the underside, since this will be better preserved. If the water above the substrate was oxygenated then scavengers may have disrupted, and encrusters (especially oysters) obscured, the skeleton. In shelly layers skeletons are likely to be somewhat dissociated because of reworking and/or hydraulic activity and scavenging. In bituminous mudrocks isolate bones have probably dropped from decomposing, floating carcasses. Preservation within calcareous concretions will necessitate acid preparation in the laboratory. Often the skeleton extends beyond the concretion. Look out for unusual aspects such as evidence of skin (may be no more than a colour change or film beyond the skeleton), or stomach contents, or unborn juveniles.

Ostracoderm and other Devonian fish have large and often massive bony plates for part of their armour, which were highly resistant to attrition. They are a common component of fluvial gravel lags. Complete skeletons can be found in flood slump deposits. But, under the acidic conditions of a peat bog, for instance, enamel is slowly decalcified, leaving only dentine. Similarly, bone can become decalcified and 'soft'. In organic-rich mudrocks bone becomes compressed and distorted. The cartilagenous skeleton of sharks preserves less readily, though may become replaced. In Mesozoic red beds calcitic egg shells can be common.

4.4   Trace fossils (W; including coprolites, faecal pellets and faecal strings)

Trace fossil taphonomy may be considered in exactly the same way as the taphonomy of body fossils: formation, stratinomy and diagenesis (Fig. 4.32).

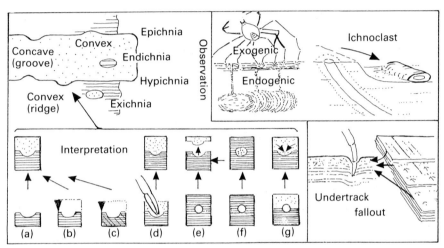

Figure 4.32   Modes of preservation of trace fossils. (*Left*) Terms for preservational description (e.g. convex epichnia, or epichnial ridge, (singular epichnion), and seven possible interpretations of a convex hypichnion (hypichnial ridge): (a) exogenic-formed groove (above substrate); (b) groove made on a degraded firm substrate; (c) groove cut, or boring dissolved, in lithified sediment; (d) Interface undertrack groove; (e) open burrow in mud degraded to expose groove; (f) filled burrow, in sediment subsequently degraded and groove exhumed; (g) burrow formed at interface sand-mud, with collapse of sand. (*Upper right*) Distinction between exogenic and endogenic formation, and ichnoclast. Endogenic burrow represented by infaunal echinoid producing meniscate backfill. (*Lower right*) Formation of undertracks (*see* Fig. 4.33)

For trace fossils in palaeoecology see §6.2.4, in sedimentology see §7.3.

A. Relation to substrate (stratinomy)

Basic questions

1. Are you looking at the actual sediment/water/air interface on which the trace was formed? (Note: the majority of subaqueous traces are today formed at or above the substrate surface (exogenic), but the vast majority of fossil traces were endogenic.) Or,

Criteria

Ridge of displaced sediment to either side, mound of sediment behind a foot or claw print.

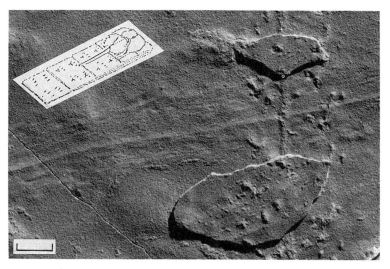

Figure 4.33 Undertrack preservation of limulid track viewed from lower surface. The laminated siltstone has split along two levels, revealing a lower level with traces of anterior appendages, and a slightly higher level with additional traces due to the pushers and telson. The telson trace cuts a set of traces made from a shallow depth and below the substrate surface used by the first animal (×1.0) (Parrsboro Formation, Nova Scotia, Upper Carboniferous). Scale bar = 1 cm

2. Are you looking at a parting below the surface, bearing marks of claws etc. that penetrated the substrate surface i.e. undertracks? Analyse the differences between each successive layer. Or,

An undertrack fall-out will be evident on successive layers. (Fig. 4.33)

3. Are you looking at a trace made at the interface of two lithologies, or a burrow within a bed? Either could have formed well below the substrate surface colonized. In the case of *Ophiomorpha* this could have been up to 4 m. Or,

(Fig. 4.32)

4. Can you identify where the substrate surface actually was? Or,

This is seldom possible. Undertracks are OK. Funnel openings and faecal castings are rare. Shallow-tier traces (Figs 6.23, 6.24, 7.13, 7.23) such as *Cruziana*, *Rhizocorallium*, *Planolites* were mostly formed at less than 2 cm depth.

5. Was the original substrate surface penecontemporaneously eroded? Or,

Termination of burrow against erosion surface.

6. Was the interface trace the result of penecontemporaneous washing out of burrows?

(*See* Figs 4.32e, 7.11, 7.12)

For sedimentological analysis it is often sufficient to describe the stratinomical relationship (Fig. 4.32). The lower figure gives an indication of possible interpretations.

B. *What was the consistency of the sediment at the time of formation of the trace (see §7.3.3)?*

Observations

Deductions

1. In mudrocks, sediment highly diffuse, mottlings, virtually homogenous.

More or less fluid sediment.

2. Most burrows lined.

Sediment soft/loose.

3. Compressed, collapsed burrows in sand.

Sediment loose to cohesive.

4. Deformed crypts, deformed vertical burrows.

Sediment cohesive.

5. Cylindrical burrows in sand.

Sediment firm to cohesive.

6. Borings truncate shells, ooids and matrix evenly.

Sediment lithified.

Vertebrate footprints are sharpest when they were formed in damp, slightly muddy sand which, when dry, was able to resist penecontemporaneous erosion. If a trackway is followed, it may well show changes indicative of wetter or softer sediment. Large vertebrates often impressed deeply into the sediment, possibly deforming the sediment below. The print may have been irregularly filled. The imprint immediately below the foot bears the closest resemblance to the actual foot, other partings to a lesser degree.

C. *How has the trace been affected by taphonomic processes (penecontemporaneous erosion, compaction, lithification)?*

The relatively unconsolidated fill of a burrow cut into consolidated sediment tends to channel migrating pore fluids. The wall of a burrow cut into loose sediment and stabilized by organic material, with or without pelleted lining or impregnation, again affects pore fluids. All such burrows act as potential foci

Figure 4.34 Ammonite in phosphatic concretion from Gault Clay, Kent, Lower Cretaceous. Specimen indicates at least three stages of phosphate growth, probably involving repeated burial and exhumation, and indicated by differences in colour of the phosphate. Scale bar = 1 cm

of mineralogical change and concretion formation, either around or within the burrow, e.g. pyritized, calcified, or silicified burrow fills, glauconitic linings. Diagenesis generally enhances bioturbation structures.

This section is, of course, principally concerned with distinct trace fossils. Where the process of bioturbation has been more intense and earlier burrows indistinct, the same principles apply, but are less readily resolved in the field, and it is generally necessary to collect oriented blocks for later slabbing (and possibly X-radiographic treatment) to determine stratinomic- and interrelationships (ordering and tiering; §7.3.6).

4.5   Concretions
(Figs 4.8, 4.9, 4.10, 4.28, 4.29, 4.34; see also §4.3.1)

It is always a joy to find a complete fossil encased in a concretion (nodule). The aspect that arouses most enthusiasm is the three-dimensional form of the scarcely compacted specimen, particularly if material from the adjacent sediment is strongly crushed. But there is a price to be paid: the difficulty of extracting the specimen(s)! Always try and split a concretion parallel to bedding. Keep counterparts, especially if the rock has split within the skeleton. In mudrocks the amount of compaction that the fossil has suffered will relate to the time of formation of the concretion. Septarian concretions can often be disappointing because of extensional cracking (which may also pass through the skeleton). The earlier the concretion formed, the more likely it is to have enclosed organic parts that are normally lost, e.g. soft tissue (§4.3.1), ammonite jaws and radula. Spherical concretions will generally be earlier than flattened ones (unless the flattening was controlled by lithological change, as with a concretion in a mudrock just below a sandstone).

Impressions in primary flint and chert often display an unusual amount of information. For instance, echinoid spines extracted from white chalk are generally imperfectly preserved, however careful preparation might be. A latex pull of the flint impression of a spine however displays the delicate thorns with perfect form. Aragonite on the other hand generally dissolved prior to the formation of flint (Fig. 4.29). Flints often enclose a hollow left by a dissolved sponge. Cracking these reveals flint meal (dusty sponge spicules, foraminifera and ostracodes – the latter partly replaced by silica). The flint meal biota is often richer and more diverse than the surrounding chalk.

Concretions of all types may be reworked intraformationally, or as pebbles and cobbles, into younger sediments where they may act as hard substrates for colonization. Intraformational reworking can be important in stratigraphy (Fig. 4.34).

4.6   Dolomite replacement and silicification

Most of the modes of preservation referred to above concern mineralogical change to the actual skeleton (replacement with a void stage, or neomorphism). Whatever has happened to the matrix has been independent of the skeleton. Wholesale replacement of the rock (skeleton and matrix) causes problems. Do not necessarily be put off by the common use of dolomite or chert in the formation name, e.g. Langland Dolomite. Examination of the rock with a hand lens may reveal partial dolomitization with scattered rhombs only. Dolomitization or silicification may be local and affect different groups differently. Echinoderm material generally resists dolomitization more than other groups. Silicification may be obvious in the field, or cryptic. Calcitic rugose corals, brachiopods and oysters are quite commonly coated by chalcedony. Wholesale silicification of the fauna in a limestone may not be easy to recognize on freshly broken surfaces. Look carefully at weathered surfaces. In certain limestones trilobites, ostracodes and bryozoans have been silicified, but the brachiopods remain calcitic. Use a hand lens to observe acid reaction.

4.7   Deformation of fossils

Deformation of fossils is due to two processes (Fig. 4.1). Compactional deformation is followed by tectonic deformation readily distinguished by the structural state, cleavage, folding, etc. of the rock. Any deformation affects recognition, identification and reconstruction. This is particularly true of groups without a substantial mineralized skeleton: plants, graptolites and many ammonoids.

Fortunately most fossils do not spread on compression (because of confining pressures) but crumple. Thus, lateral dimensions are virtually unchanged. Exceptions are relatively unsupported parts, e.g. apertures of graptolites and cephalopod body chambers, which fracture and push into the matrix. This is most likely to occur in very soft sediment. Uneven fracture and spread is generally obvious. Factors that lead to variation in form of the flattened fossil are decay distortion, different attitudes of appendages and, especially, variation in the original attitude of the organism to the bedding.

Reconstruction of compressed fossils will generally be undertaken in the laboratory. The restoration technique described by Briggs and Williams (1981) is useful. This involves photography of models and comparing the photographs with the compressed material. Silt, sand and coarser grade sediment scarcely compacts unless rich in organic (plant) matter. The effects of pressure dissolution are generally obvious in limestones but are less obvious in mouldic preservation in sandstones or siltstones.

There are sedimentological implications with compacted shells. For instance, fractured shells in a concretion may provide an indication of the burial depth at which lithification took place, if the loading threshold for shell fracture can be determined. Not enough data is yet available however to provide useful information for the field.

Check whether or not early dissolution took place, and do not confuse septarian cracks (tensional) with load fracture. The presence of intact shells in muddy sediment means that either the load has been low or that the enveloping sediment became lithified relatively early. Broken shells show that breaking pressure was reached before lithification.

A provisional scale of natural breakage (from field observations) under compaction for some fossils (after Brenner and Einsele, 1976) is presented below:

Increasing
resistance

Jurassic ammonites: *Lytoceras,*
Phylloceras, Harpoceras (but *see* §4.3.2)
Pseudomonotis, Inoceramus
Posidonia
Triassic ammonites: *Ceratites*
Terebratulids, rhychonellids
Gastropods
Crinoid ossicles

A scale of breakage of common modern bivalves:

Increasing
resistance

Pholas
Mya, Mactra
Mytilus
Pecten
Cardium

References

Most palaeontological and palaeobotanical texts (*see* Chapter 6) have sections on taphonomy and fossil preservation. Useful, but mostly detailed and specific, references to the 'state-of-the-art' are included below, together with references cited.

AGER D V 1963 *Principles of Paleoecology* McGraw Hill, 371 pp

ALLISON P A 1988 Soft-bodied animals in the fossil record: the role of decay in fragmentation during transport. *Geology* 14: 979–81

BRENNER K, EINSELE G 1976 Schalenbruch im Experiment. *Zentralblatt für Geologie und Paläontologie* Jahrgang p. 349–54

BRETT C E, BAIRD G C 1986 Comparative taphonomy: a key to paleoenvironmental interpretation based on fossil preservation. *Palaios* 1: 207–27

BRIGGS D E G, WILLIAMS S H 1981 The restoration of flattened fossils. *Lethaia* 14: 157–64

EGLINGTON G, CURTIS C D, MCKENZIE D P, MURCHISON D G (Eds) 1985 Geochemistry of buried sediments. *Philosophical Transactions of the Royal Society of London* A 315 233 pp. (A collection of papers relating to marine and freshwater diagenesis)

HUTTON A C 1986 Classification of Australian Oil Shales. *Energy Exploration & Exploitation* 4: 81–93

MARTILL D M 1988 Preservation of fish in the Cretaceous Santona Formation of Brazil. *Palaeontology* 31: 1–18

REX G 1983 The compression state of preservation of Carboniferous lepidodendroid leaves. *Review of Palaeobotany and Palynology* **39**: 65–85

REX G 1985 A laboratory flume investigation of the formation of fossil stem fills. *Sedimentology* **32**: 245–55

REX G, CHALONER W G 1983 The experimental formation of plant compression fossils. *Palaeontology* **26**: 231–52

SEILACHER A 1971 Preservational history of ceratite shells. *Palaeontology* **14**: 16–21

SEILACHER A 1976 Preservational history of compressed Jurassic ammonites from southern Germany. *Neues Jahrbuch für Geologie und Paläontologie Abhandlungen* **152**: 307–56

SMITH A 1984 *Echinoid Palaeobiology* George Allen & Unwin, 190 pp

SPICER R A 1981 The sorting and deposition of allochthonous plant material in a modern environment at Silwood Lake, South Park, Bracknell, England. *United States Geological Survey, Professional Paper* **1143**, 77 pp

STACH E (Ed) 1982 *Stach's Textbook of Coal Petrology*. Gebrüder Borntraeger, 535 pp

Two journals have special issues on taphonomy:
Palaios 1986 volume **1** (3)
Palaeogeography, Palaeoclimatology, Palaeoecology 1988, volume **63**

5

Pseudofossils, and stratigraphical and structural errors

Two-thirds of Earth's history is represented by Precambrian rocks, and it is only towards the top of the Precambrian that evidence of larger organisms (some definitely metazoans) and of metazoan activity (burrows), can be recognized. There is, though, abundant evidence of simple plants extending far back into the Precambrian. The temptation to be the first to find the oldest metazoan is strong, but it is wise to be aware of possible pitfalls, and the criteria to use on structures of uncertain origin. While this applies particularly to Precambrian sediments, it is also relevant to Phanerozoic sediments, since the same sorts of problem occur there. There are three categories of false lead:

1. *Activities of modern animals* For example, burrows in Precambrian sediments which have subsequently been found to be the work of modern termites probing deep to get to the water table, or burrows, also in Precambrian sediments, which were later recognized as due to modern annelids, penetrating into joints and slightly dissolving the calcareous sediment. The clue is to look carefully at the bedding relationships (§4.4, 6.2.4).

2. *Structural or sedimentological errors* For example, in a belt of Precambrian quartzites a unit of down-faulted Tertiary sandstones with burrows was initially unrecognized as such. In Siberia, in late Precambrian dolomites, a layer of fossils was unrecognized as the infill of a sedimentary (Neptunian) dyke system from the Lower Cambrian (§7.1).

3. *Pseudofossils* Rock structures that resemble fossils, that look like plants or animals or parts of these, or look like the activities of animals. It is important to pose the right questions to decide whether a structure is a fossil or a pseudofossil. First consider taphonomic aspects and the organization of sedimentary structures (bedding, lamination). There is now a long list of structures that resemble animals, animal activity or plants to which a Linnean name was applied, but which have now been shown to be of inorganic origin (Häntzschel, 1975).

It is generally easier to try to find a sedimentological solution than to obtain proof of biological origin. Refer to the boxes on Fig. 5.1. Gas or water escape

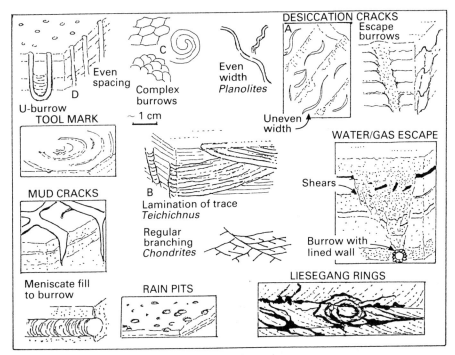

Figure 5.1 Field evidence that might be used to distinguish between trace fossils (unboxed) and some sedimentary structures (boxed). (A) desiccation cracks in mud ripple troughs; (B) lamination formed by the upward migration of the *Teichichnus*-animal, which may sometimes resemble small-scale cross-lamination; (C) complex burrows: see §6.2.4 and §7.3; (D) U-burrow (Fig. 7.18) and vertical single burrows

structures often look like burrows, but an upward cone with sheared walls is not the structure produced by an animal moving upwards. If the sediment is loose, confirmation of an inorganic origin can often be found by probing to the base of the cone to a ruptured burrow. Sinuous, tapering desiccation cracks in ripple troughs (the so-called *Manchuriophycus*), or polygonal V-shaped desiccation cracks, likewise differ from the even width of a burrow. The irregularity in size of rain pits distinguishes them from the more uniform width of burrow openings. Here, try and locate the burrows extending downwards normal to bedding. Certain tool marks (§7.4) are probably due to wave action on attached weed. Liesegang rings or bands, secondary structures caused by rhythmic precipitation (generally of iron oxide) in a fluid-saturated rock, often resemble burrows. Primary sedimentary structures such as cross-lamination passing across the rings demonstrate their secondary origin. Similarly, primary lamination passing across concretions is proof of their stratinomical relationship. But note that concretions which have a bodyfossil-like shape (Fig. 5.2) may nevertheless encase true fossils (§4.5), or have formed along burrow fills. Dendritic markings (Fig. 5.2) caused by deposition of a manganese film along bedding or joint surfaces are another common pseudofossil. Diagenetic structures usually give themselves away because of the mineralogy (pyrite, flint etc.).

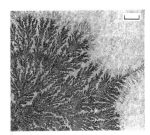

Figure 5.2 (Below) Banded flint suggesting annelid segmentation, Upper Cretaceous. (Above) Dendritic markings suggesting moss, due to manganese oxide on joint plane.
Scale bar = 1 cm

One should use one's knowledge of evolution with confidence. For example, apparent fragments of an eurypterid in positively-identified Precambrian rocks, cannot derive from an eurypterid. Closer inspection would soon reveal the absence of scales on the supposed carapace plates, which turn out to be intraformational mud chips.

For macrostructures, highly evolved structures and growth rings (ontogeny) provide evidence of biological origin. Simple symmetry and abundance are not acceptable criteria. A category, dubiofossils, has been proposed to cover such cases of doubtful origin.

There are also some true organic structures that often look inorganic, such as stromatolites and oncolites (§4.2.3). It is often exceedingly difficult to go further than indicate a strong similarity with modern examples. Note that in most cases interpretation is dependent on slabbing.

References

EKDALE A A, BROMLEY R G, PEMBERTON S G 1984 *Ichnology* S.E.P.M. Short Course notes No. 15, 317 pp (Discusses recognition of trace fossils)

HÄNTZSCHEL W 1975 Trace fossils and problematica. *In* Teichert C Ed *Treatise on Invertebrate Paleontology* Lawrence Kansas, Geological Society of America and Kansas University Press, Part W, 269 pp

6

Fossils for the palaeontologist and palaeoecologist

'Every fossil was once a living organism'

Palaeontology aims to determine the origin and evolution of the biosphere. This means recognizing and describing each taxon, knowing its origins, geographical distribution, relationships with other taxa and its demise. That much of the story is lost, because of the incompleteness of the fossil record, is irrelevant. The quest for the patterns of evolution is there. The processes of evolution are best left to the biologists.

Every year sees significant new finds made by geologists at all stages of training and levels of experience. The sort of information that can be unearthed includes: recognition of new taxa; better information on the morphology of poorly known taxa; information on geographical distribution; a new mode of preservation that allows better understanding of biological affinities; specimens showing new information on physiological functions; traces that show information about behaviour; more precise information on stratigraphic occurrence.

The aim of this chapter is to encourage the search for new evidence, as well as to indicate how to undertake field investigations, make a census, or palaeoecological analysis.

It is useful to separate the ecological and systematic aspects of fossil material. The ecological use of body and trace fossils is of wider interest to geologists, and tries to answer the questions (Table 6.1): how, where, with what?

Table 6.1 Basic questions to be asked for each fossil

To WHAT group does it belong?	(systematic assignment)
WHEN was it a living organism (or part of, or associated with a living organism)?	(time aspect)
WHENCE did it evolve?	(ancestry)
WHITHER did it evolve?	(descendants)
WHO (if anyone) has named and described it?	(taxonomy)
HOW did it live?	(palaeophysiology)
WHERE did it live?	(habitat)
WITH WHAT was it associated?	(community analysis)

For the systematist concerned primarily with morphology and evolution, the questions are: what, when, whence, whither? In the field one is concerned with recognizing just what the material is, from phylum level right down to the species. This means having a pocket field identification volume (Appendix B).

Many consider that palaeoecology can only contribute to palaeoenvironmental interpretation, and that it is not really feasible to attempt to determine what the productivity and energetics might have been for ancient sediments. But, there are a number of fossiliferous sites where the preservation is so good, that reasonable figures could be obtained (*see* Powell & Stanton, 1985). Even shell beds that are clearly allochthonous may, in certain cases, be regarded as time-averaged and representative of the environment in the long term.

6.1   Palaeontological and palaeophysiological analysis

Palaeontological analysis of morphology, ontogeny, functional morphology and evolution requires particular types of evidence: completeness of individuals, perfection of detail and often original shell structure. The classification of fossiliferous sediments (Chapter 1) gives an indication of the most rewarding types of sediment (see also the section on fossil ores; §6.3).

Try to answer the question: what mode of preservation will be most useful in providing the information required (Chapter 4)? In every case one is collecting for subsequent preparation and examination. Material may be limited, as with small pockets in a scourfill structure, or just the exceptional 'lucky' occurrence. It is important to make and record observations about the site before extracting any fossils. Sketches and measurements of orientations may prove useful later. A graphic log and photographs may also be useful (§3.4). For evolutionary studies, sufficient numbers of particular taxa, and of different growth stages, are required (§3.5). In evolutionary studies, it is important to get the stratigraphy absolutely right. For instance, pose the questions: is there a strike fault that repeats the stratigraphy, or is there a stratigraphic break in the section? Record on log sheets or in a notebook the exact location of samples collected. If a change in the biota is spotted through a sequence that might be attributable to evolutionary processes, then it is always worthwhile considering the possible evolutionary pattern (punctuated equilibrium or phyletic gradualism) and the process (isolating mechanisms or change in timing of ontogenetic events – heterochrony) that was involved.

In palaeophysiological analysis one is trying to determine just how a fossil 'lived', how it once fed and on what, how it respired, grew, reproduced, moved and protected itself. Methods of feeding used by modern animals are shown in Appendix D. Skeletal morphology is influenced by several factors (Fig. 6.1):

1. Phylogenetic factors e.g. the inherited calcitic shell of the brachiopod and its accretionary growth.

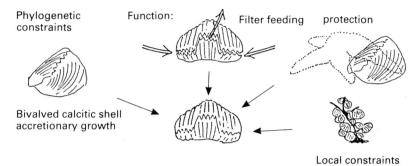

Phylogenetic constraints

Function: Filter feeding   protection

Bivalved calcitic shell accretionary growth

Local constraints

Figure 6.1   Major factors influencing skeletal morphology, as illustrated by a bivalved, calcitic shell with accretionary growth: *Rhynchonella*

2. A 'genetic pull' between two or more functions during evolution e.g. feeding and protection.
3. The design of an efficient mechanism for a particular function; e.g. for filter feeding, (a) a 'sieve' (slender gape of valves and the lophophore within), (b) a current (produced by cilia on the lophophore), and (c) separation of inhalant and exhalant currents (the fold and sinus).

In addition, local factors may influence development of the perfect design, e.g. crowding. The 'specifications' for some physiological processes are:

Grazing (limited food resources)
Guidance system to cover maximum area with minimum effort

Predation (for most predators)
Effective means to locate, capture, kill and ingest prey

Protection (for shells)
Envelope that is strong and rigid, that completely encloses soft parts and does not interfere with other vital functions

Respiration (gaseous exchange)
External exchange system (gills, lungs, tube feet)
Circulation system; provision to prevent recirculation/choking

Reproductive mechanisms
Release of gametes so that adequate fertilization takes place under conditions for adequate dispersal and minimal loss

To satisfy oneself about the function of any structure, consider the specification for the function in mind, and how well the fossil structure meets this. (More details about functional morphology may be found in texts, e.g. Boardman *et al.*, 1987; Clarkson, 1986.) Perhaps the most sophisticated example of revitalization is Paul and Bockerlie's (1983) analysis of a cystoid echinoderm. Note that trace fossils can provide information about morphology, e.g. trilobite tracks and trilobite dactylus.

Sexual dimorphism and polymorphism are now recognized in many groups of fossils. For most groups they can be substantiated only by demonstrating the close similarity between the early stages of development and identical stratigraphic range. Similar changes in size of two 'species' through a succession, may be a clue to their sexual relationship, but check that similar changes do not occur in other taxa (i.e. that the change is ecological). Dimorphism and polymorphism also result from the alternation of sexual and asexual reproduction, but this will be evident in the field only in some larger foraminifera. Mature, 'adult' ammonites display uncoiling of the umbilical seam, modification of ribbing near to aperture (e.g. lappet), and close spacing and modification of the last septal sutures. If stunting (arrested or hindered growth) is suspected in a fauna, the only way to prove this is by (a) growth analysis: plotting growth-line spacing against increments in size in the suspect suite and in a normal suite; (b) identification of mature individuals of small size. Likely causes of stunting are low salinity (§6.2.3.4), oxygen deficiency (§6.2.3.5) and food deficiency. Much has been done on tree ring analysis as an indicator of ancient climate, and growth banding is present in many invertebrate groups, and has been established to have a definite periodicity in corals (Fig. 1.2), and several groups of the Bivalvia. Little palaeontological work has been done in this area as yet, but there is potential for the determination of life-spans and carbonate production and clues about sedimentation rate.

6.2   Palaeoecological analysis

'An ability to identify organisms accurately is an essential prerequisite for most ecological studies' – and for palaeoecology!

A full palaeoecological analysis of any sedimentary unit is a demanding task. Much of the information needed can be gleaned by careful observation without removing fossils or sediment. After all, what is required is to determine the relationship between the fossil and the associated sediment, and the relationships between the various taxa. Observations and deductions made on each taxon (autecology) or significant morphological features will be synthesized later (synecology). But, palaeoecology is concerned with more than just fossils. If ecology is the study of the interactions of living organisms with one another and with the physical environment, then palaeoecology concerns the same interactions that existed in ancient environments. The problem is to identify what the ancient environments were. All there is to go on are the sediments (facies) and the fossils. Thus, to be successful with palaeoecology a good basis in sedimentology as well as in biology and chemistry is required, i.e. an integrated approach.

The first question to ask is whether or not the fossil is in the position in which it lived, i.e. is it autochthonous? (For a planktonic or nektonic organism the questions are, how far has it drifted and has it been reworked?) If not, has it been either disturbed (parautochthonous) or winnowed, or, has it been substantially shifted and is allochthonous? In most cases the answer is clear, but it is worth calling to mind the complexities of a modern tropical reef; for instance, one with a substantial coral framework, but with patches of shell and coral sand that have formed by hydraulic action across and through the reef. A contrasting situation is the shell bed formed by a storm that has drawn numbers of starfish and crab scavengers to feed on the dead and dying. Or, consider the situation where the top of a storm-generated shell bed becomes a substrate for colonization. Any accumulation may grow by successive destructive and constructive events – a situation analagous to a human settlement (a tell).

Many organisms are able to survive displacement and skeletal damage during transport. This is particularly true of many calcareous algae, where fragments of the plants may continue to grow in their new habitat. The effect of transport with such algae is a bit like pruning a hedge: it results in an abundance of new 'shoots' to produce relatively dense rhodoliths (red algae) or oncolite (blue-green algae).

Allochthonous shell accumulations have been extensively used in palaeoecology. This is largely justified by arguing that for a particular facies, the shell beds are found to be compositionally similar. But to establish this requires careful analysis of the hydraulic processes involved, and analysis and comparison with the composition of adjacent facies, e.g. how the shelly composition of mudstones between storm layers compares with the shelly composition of the storm layer (see §6.2.3.8).

6.2.1   Autochthony, parautochthony and allochthony

Criteria to determine autochthony and parautochthony are listed below. When examining a fossil occurrence, it is important at the outset to look carefully at the hydraulic and sedimentary regime that might have pertained, and to consider whether this regime would have allowed colonization to take place and to be sustained. In the case of a thick turbidite or ash for instance the speed at which the sediment was deposited means that organisms could only have colonized the top of the bed, though such an 'event' bed may have smothered an earlier community. The best way to determine the sedimentary regime is to examine the stratification (if necessary by cutting and partly polishing a section at right angles to the bedding) and to look for evidence of 'event' beds or of bioturbation. The inferences that can be made from such information are described in detail in §7.3.

The following criteria should be used to determine autochthony and parautochthony:

1. *Normal life attitude* Bivalve shells (Fig. 6.2), especially infaunal species, are commonly found in life position, as are corals and other substrate

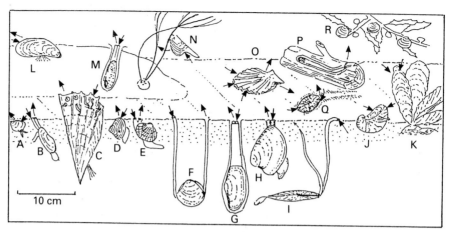

Figure 6.2 Bivalve life attitudes: (A) *Nucula* (nut-shell); (B) *Yoldia*; (C) *Pinna* (fan-mussel), with encrusting serpulids; (D) *Astarte*; (E) *Cerastoderma* [*Cardium*] (cockle); (F) *Lucina*; (G) *Mya* (sand-gaper); (H) *Mercenaria* (guahog); (I) *Tellina*; (J) *Gryphaea* (Devil's toe-nail) (secondary soft-bottom dweller); (K) *Crassostrea* (American oyster); (L) *Mytilus* (mussel); (M) *Pholas* (piddock); (N) *Pteria* (wing-oyster); (O) *Pecten* (scallop); (P) *Teredo* (shipworm); (Q) *Glycymeris* (dog-cockle); (R) *Posidonia, Bositra* (?epiplanktonic, ?nektoplanktonic). Arrows indicate incurrents and excurrents

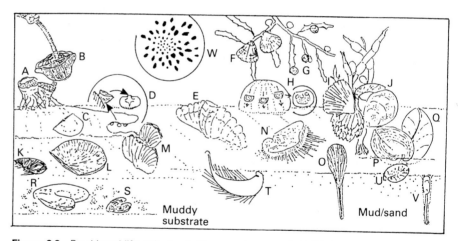

Figure 6.3 Brachiopod life attitudes (with approximate scales): (A) *Prorichthofenia* (coral-like, umbonal cementation, plus spines) ×0.3; (B) *Linoproductus* (clasping) ×0.3; (C) *Irboskites* (cemented) ×0.5; (D) *Morrelina* (cemented, photonegative thecideidine, brachial valve shown) ×2.0; (E) *Spirifer* ×0.5; (F) *Rhynchonella* (thin shelled, epiplanktonic?) ×1.0; (G) epiplanktonic inarticulate ×1.0; (H) *Crania* (ventral valve shown, cemented to echinoid test) ×1.0; (I) *Rhynchonella* (pedically attached to algae) ×0.5; (J) terebratulid (pedically attached) ×0.5; (K) spine-stabilized chonetid ×0.3; (L) strophomenid (recliner) ×0.3; (M) orthids (pedically attached) ×0.3; (N) *Waagenoconcha* ('snow-shoe' spines) ×0.3; (O) *Terebrirostra* (semi-infaunal) ×0.3; (P,Q) *Conchidium, Pentamerus* (unattached, with umbonal burial) ×0.25; (R,S) *Pygites* ×0.5, *Dicoelosia* ×1.0 (current flow maximization); (T) quasi-infaunal productid ×0.5; (U) free-living terebratulid with thickened umbos) ×0.3; (V) infaunal lingulid ×0.2; (W) *Podichnus* (etch trace of divided, root-like pedicle) ×12.5

encrusting groups. Look out for pelmatozoan holdfasts or 'rooting' structures. Nestling echinoids between coral branches may be identified by radially arranged spines. Brachiopods (Fig. 6.3) are mainly epifaunal and hence less frequently found in life attitude. The life attitude of many fossil brachiopods is therefore uncertain.

2. *Size frequency distribution* This (Figs 2.12, 6.18) would seem to offer clear evidence about authochthony; but there are many factors such as winnowing, additions, spatfall frequency and growth rate, which make it difficult to apply. An occurrence of articulated shells of all sizes is likely to be autochthonous (*see also* §6.2.3.10).

3. *Completeness and sorting* Leaves and fronds of complex and delicate morphology cannot have been transported far. The extent of skeletal dissociation and valve disarticulation can also be useful indicators of autochthony. Collect for measurement, or measure frequencies of left and right valves, cephala, pygidia etc. Always consider that large and massive skeletons require the application of considerable force to be shifted. Interpretation of the data can be difficult because, for example, varieties of brachiopod resist disarticulation differently: those with strong tooth–socket structure tending to remain articulated. There may also be considerable differences in the breakage potential between valves, as between some pedicle and brachial valves. The number of right valves of oysters is always reduced compared with the number of left valves. Incidentally, free left valves must have been attached to something. It may be possible to identify what this was by looking at the preserved sculpture on the attachment surface. How close are the results to expected frequencies (Appendix C)? With more severe disturbance consider what sorting and winnowing has taken place, and under what type of hydraulic regime and substrate conditions. Sorting is largely influenced by size, composition and shape. For instance, different skeletal elements of a trilobite behave differently under flow.

4. *Shell breakage and damage* The bulk of shell breakage can be attributed to three causes:

   (a) Breakage by hydraulic action prior to deposition. This is of little or no ecological significance, since breakage relates to the form and microstructure of the shell.
   (b) Breakage due to compaction (§4.7).
   (c) Breakage due to weathering effects.

The two latter causes should be recognizable, since it will be possible to reassemble fragments as they are extracted from the sediment. In younger sediments, there may be other causes of shell breakage: predation (shell chipping) by crabs (Fig. 6.5) and predation by birds (probably the main cause of shell breakage on modern tidal flats). Many demersal fish, shell-eating sharks and large crustaceans can fragment shells but there is no recognizable pattern of breakage.

5. *Attitude and preferred orientation*   Plotting valve attitude (Fig. 2.12), convex up, concave up, oblique, or perpendicular may give an indication of the hydraulic processes involved (§7.2.2). Consider the likely original attitude of the components. Low degree of preferred attitude (associated with similar frequencies of valves) would suggest little hydraulic disturbance. For gastropods, absence of preferred orientation in the horizontal plane is the best indication of parautochthony, but overall random orientation and attitude is probably due to bioturbation. Check that the fabric supports this inference.

Organisms respond to a number of orienting influences during life, particularly to light, gravity, currents and food resources. Burrows along muddy (and more nutritious) ripple troughs, and aligned suspension feeders are common in the fossil record.

6. *Three-dimensional clustering*   This can be due to a number of processes. Local pods of shells may be related to burrow infills, gutters or other such structures, load structures, or a portion of a shell bed remaining following erosion. Where the cluster comprises but a few species with many individuals with articulated valves, and some valves deformed because of crowding, then one must have a natural cluster (cf. Fig. 6.4) that was overwhelmed by sediment. Laterally-extensive clusters no more than a few cm thick may represent local, firm or hardground colonization and may have a very diverse biota. A 'cluster' more or less confined by a concretion (dogger) probably represents early cementation-retaining shells, but diagenetic loss of laterally adjacent shells (Fig. 3.1).

7. *Anomalous presence/absence of selective elements*   Are there any elements that might be expected, but are conspicuously missing because of suspected selective hydraulic removal, or because of selective skeletal dissolution, e.g. echinoids on an oyster patch where there are indications of the tooth marks (grazing activities) of the echinoids. Or, is there the anomalous presence of byssate and encrusting forms on soft substrata?

8. Pose the questions: how does the faunal composition of the bed compare with that in adjacent lithologies? Could the fossils have been winnowed from the sediment below?

9. Consider the extent and distribution of encrusters and whether their orientation, with respect to the substrate (shell) indicates encrustation during the life of the host (epizoan) (Fig. 6.15), or encrustation of a dead shell (epilith). If the latter, has there been subsequent reworking and turning? Encrustation of infaunal bivalves at only the posterior part of the shell may indicate that this portion of the shell was exposed during life (Fig. 6.2c), or that penecontemporaneous erosion had partially exposed the shell. Shells with encrusters on both inside and

Figure 6.4 Terebratulids attached to a gastropod columella (Recent, Australia). Scale bar = 1 cm

Table 6.2 Good criteria for parautochthony in marine mudrocks i.e. where autochthony uncertain

1. Matrix-supported shell fabric
2. High degree of bioturbation
3. Similar frequencies of each valve (in bivalves) even if hydraulically distinct
4. Local winnowed horizons
5. Many shells convex down, and nesting common
6. Little or no preferred orientation
7. Lack of abrasion
8. Lack of encrusters or borers on infaunal shells

Figure 6.5 Gastropod (*Sycostoma bulbus*) chipped by a crab (Barton Formation, Middle Eocene, Highcliffe, southern England). Scale bar = 1 cm

outside must have been on the substrate surface for some time. But are all the shells similarly encrusted? There may be evidence of several phases of reworking.

10. Decide whether the sediment is matrix or skeleton supported. The former is difficult to explain by hydraulic processes (unless one is clearly dealing with a turbiditic bed, or there has been extensive bioturbation).

11. Applications of the above criteria to some typical shelly beds are shown in Figs 1.2 and 6.10. Beds analysed in Fig. 6.10 show no shells in life position but varying degrees of dissociation, preferred orientation and left/right valve ratios. Diversity can also be an indication when taken into consideration with other evidence.

6.2.2 Buildup-producing organisms and their roles

Buildup-producing organisms are diverse, have evolved and been repeatedly replaced over geological time (Fig. 3.4). Many are difficult to identify, especially in the field. Their roles in forming buildups and in producing loose sediment are summarized in Table 6.3, and the relationship of growth form

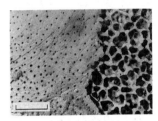

Figure 6.6 Scallop shell (*Pecten*) partly split to reveal cavity system (*Entobia*) caused by sponge *Cliona* (Recent, Newfoundland). Scale bar = 1 cm

Table 6.3 Roles of buildup-producing organisms

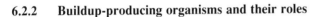

1. *Binding, encrusting and stabilizing*
   (closely attached, ± cemented, roots, filamentous, mucilagenous)
   Major : laminar rhodophytes, tabulate corals, blue green algae, oysters, poriferans
   Minor : stoloniferous bryozoans, tabulate corals, stoloniferous foraminifera, crinoid holdfasts, sclerosponges, vermetid gastropods

2. *Frame builders*
   Primary and macro-sized: lamellar, massive, tabulate corals; strongly branched rugose and scleractinian corals; some bryozoans; rudistid bivalves; some sponges; stromatoporoids; rhodophytes, tubiphytes

   Secondary (without ability to construct solid frame but infilling cavities etc.) as above but smaller scale, diverse groups especially foraminifera, bryozoa, sponges

3. *Bafflers* (cylindrical ± branching, vase-form)
   Corals, large serpulids, sea grasses, mangroves, fenestellids, archaeocyathids, stalked crinoids, sponges

4. *Producers of loose sediment*
   Sand sized, free or weakly-linked elements: articulated green and red algae, crinoids, larger foraminifera, sponges, foraminifera attached to sea grasses. Silt to clay size elements: pelagic foraminifera, coccoliths, green algae

Growth form		Environment	
		Wave energy	Sedimentation
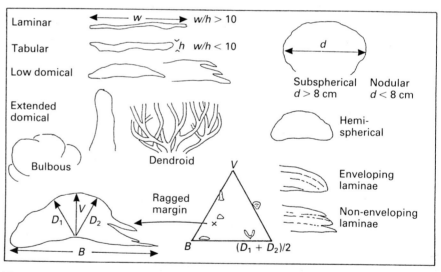	Delicate branching	Low	High
	Thin, delicate plate-like	Low	Low
	Globose, bulbose columnar	Moderate	High
	Robust, dendroid branching	Moderate	Moderate
	Hemispherical, domal, massive	Moderate/high	Low
	Encrusting	Intense	Low
	Tabular, laminar	Moderate	Low

Figure 6.7 Growth form of reef-building organisms related to wave energy and sedimentation (after James, in Scholle *et al.*, 1983)

Figure 6.8 Stromatoporoid morphology and parameterization of shapes (except for dendroid and irregular forms) (after Kershaw and Riding, 1978)

to wave energy and sedimentation rate are depicted on Fig. 6.7. The shapes of stromatoporoids, so characteristic of Silurian and Devonian buildups can be parameterized (Fig. 6.8). The geological literature often refers to distinct development stages in ancient coral and stromatoporoid buildups (Fig. 6.9), but work on Pleistocene and modern reefs indicates more complex successions (Crame, 1981; *see also* §6.2.5). In the Jurassic example described in §2.1

Stage	Limestone type	Diversity	Colony shape
Domination	Bindstone–framestone	Low to moderate	Laminar/ encrusting
Diversification	Framestone (bindstone) + Wackestone (matrix)	High	Variable: domal, massive, lamellar, branching, encrusting,
Colonization	Baffle–floatstone wackestone matrix	Low	Branching, lamellar, encrusting
Stabilization	Variable, generally coarse grained	–	–

Figure 6.9   Idealized stages in the development of a reef core, and associated type of limestone, species diversity and colony shape (after James, in Scholle *et al.*, 1983)

material has been introduced into the coralliferous unit, and adjacent grain-stones carry elements derived from the buildup. Detailed work is necessary to evaluate this 'export–import' model which Bosence and colleagues (1985) are pioneering on modern south Florida mud mounds and associated sediments.

6.2.3   Ecological factors

There are three all-important ecological gradients in the World's oceans. Firstly, there is the *depth* gradient. Light intensity (and hence photosynthetic activity), temperature, and oxygenation all decrease with depth.

Secondly there is the *shore to open ocean* gradient. Nutrient levels tend to be higher in inshore waters either because of upwelling or because of land-derived material. This means that inshore waters have far higher levels of productivity than their offshore equivalents.

Thirdly there is the *latitudinal* gradient. In low latitudes the seas receive a greater amount of both light and heat than they do towards the poles. Waters at low latitudes therefore tend to be more productive.

The above are very broad generalisations and modern seas show that local factors can disrupt expected gradients. For example, nutrient-rich waters upwelling in polar seas can cause very high levels of productivity, or sediment suspended in shallow nearshore waters can greatly reduce expected light levels. It is reasonable to assume that the first of these gradients was as important in the past as it is today, but latitudinal gradients certainly seemed to have varied in intensity. In the Mesozoic for example the temperature difference between the poles and the tropics was apparently much less than it is now. The shore-to-open ocean gradient in the Mesozoic was also probably generally more gentle because higher sea levels gave rise to extensive shallow shelf seas.

When making palaeoecological deductions based on these major environmental gradients, it is therefore useful to remember both the importance of local factors and of past sea-levels and climates.

An appreciation of the role of ecological factors in palaeoecology is best gained before going into the field. Some data are given in Appendix D. The role of many ecological factors may be estimated in ancient sediments by five methods:

1. by comparison with the ecological requirements known for modern taxa, by close comparison for Pleistocene and Tertiary taxa (at species or genus level), or by inferences made on older taxa (generally at genus or family level and with reduced degree of certainty);
2. by making inferences about various morphological features related to specific ecological conditions e.g. spinosity;
3. by making inferences from sedimentary parameters e.g. ooids, evaporites, facies relationships;
4. from samples collected it may be possible to make determinations by: chemical (including isotopic) methods for palaeotemperature and palaeosalinity;
5. utilizing faunal and floral gradients with respect to the equator, e.g. high diversity and warm temperature at low latitudes.

In general, the precision with which various factors can be determined decreases with geological age. Thus, in the Palaeozoic most determinations are rather speculative, and in the Precambrian largely unrelated to the present – anactualistic. It is important to try to use more than one method to determine any parameter. Ecological factors that are of concern to palaeoecology are shown in Table 6.4.

When an appreciation of the nature of the palaeoenvironment in question has been gained, it is useful to assess what changes in ecological factors might be expected, bed-to-bed, and laterally, across the field area (Table 6.5). This will focus attention on those factors that are of most significance for the particular situation. Any organism relates to a tolerance range and optimum value for each factor. There are also general ecological principles and 'laws' which may occasionally be useful (Table 6.6).

In the field, one's view is generally more limited so far as lateral change is concerned, and it is the vertical, stratigraphical, changes that are often of most concern. As is mentioned (§1.3) Schäfer (1972) recognized the relationship between oxygenation, benthos and stratification. This scheme can be extended to include nutrients and degree of bioturbation (Fig. 6.11).

For terrestrial plants, temperature and precipitation are the most important ecological factors, followed by light (length of day, cloudiness) and soil conditions. Indications of these factors are preserved in the plant morphology and soil. Although identification of the stomatal and other microscopical

Table 6.4 Ecological factors

Physical	Chemical	Biological
Depth	salinity	food resources
Temperature	oxygenation	associations between organisms
Light conditions	hydrogen-ion	abundance
Turbulence (see Chap. 7)	concentration	diversity
Substrate conditions		dispersion
& geomorphology		ecological succession
Rainfall		palaeogeography
Latitude		dispersal
Distance from shore (oceanity)		birth & mortality rates

Plus: temporal changes in many of the above (daily, monthly, annual etc)

Table 6.5 Ecological factors and rate of geographical and stratigraphic change

	Rate of expected change	
	Geographical (over a field area)	Stratigraphical (bed-to-bed in a section)
Climatic	slow changes only, over extensive areas and minimized during times of global equable climate	slow; unlikely over less than a stage (but rapid in interglacials)
Light	high; may be local because of turbidity. Low with depth change	rapid if associated with turbidity; slow with depth change
Substrate condition	often very localized	rapid but reflected in lithology
Depth	if eustatic, likely to affect extensive area; if tectonic, rapid and may be quite local; if magmatic, rapid and local	significance decreases with increasing depth. Can be rapid in enclosed seas
Turbulence	generally very localized	rapid e.g. tidal fluctuations, storm events
Salinity	local to regional fluctuations mostly associated with nearshore or lacustrine environments	rapid; seasonal effects may be over small thickness
Oxygen	conditions of low oxygenation can be local or very extensive	change quite rapid

Table 6.6 Some ecological principles and laws

Leibig's law	The distribution of a species is determined by that ecological factor which is present in minimum amount (If two (or three) factors approach minimum, survival is especially critical)
Allen's rule	Body extremities tend to be smaller in cold climates
Bergman's principle	Species attain larger size in colder regions
Gause's principle	Two species with the same ecological requirements cannot (normally) exist together

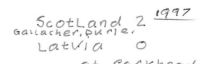

characters is a laboratory study, leaf outline (§6.2.3.2) and growth rings can be seen in the field. Paleosols are potentially useful indications of climate. Most, and of course, pre-Upper Palaeozoic soils, do not display evidence of roots (root mould, root casts or petrified roots). Look at the context of the suspect lithology for the likelihood of soil formation, and for evidence of karstification, destratification, colour mottling, calcrete, silcrete, pseudo-anticlines.

6.2.3.1 Depth and light Apart from shoreline facies, determination of palaeobathymetry can only be an estimation. In the field use the following:

1. shoreline features e.g. tidal stratification, herring-bone lamination, tidally-produced accretion sequences, tidal cycles, interference, terraced and truncated ripples, desiccation features: bird's eye structures, rainprints, footprints;
2. ooids (generally <5 m, 10–15 m); grapestones (similar but in sheltered areas);
3. oscillation ripples are generally indicative of depths above normal wave base, but always consider the whole context;
4. storm event beds (§7.3.1), associated with sedimentary structures such as hummocky cross-stratification, suggest depths to 200 m (storm base), but always consider the whole context;
5. syndepositional, early diagenetic minerals: glauconite forms today at 125–150 m, chamosite forms today at 10–50 m;
6. hermatypic corals (<100 m), larger foraminifera (<120 m);
7. trace fossil bathymetry (§6.2.4.1).

Five other approaches to palaeobathymetry can also be used in the field:

1. lateral change in facies and in benthonic assemblages;
2. lateral passage into buildups which display an ecological succession (providing time lines are available);
3. relief on unconformities and reefs, but remember to correct for compaction;
4. the application of aragonite and calcite dissolution depths (presently, respectively, between 1 and 2 km and between 3.5 and 5.5 km) depends on the absence of these minerals, e.g. aptychus limestones indicating depth greater than aragonite dissolution depth. These depths have fluctuated over geological time. In the Cretaceous it was shallower (calcite c. 1.2–2.5 km) subsequently deepening. The Jurassic calcite dissolution depth is estimated at c. 3.5 km;
5. light is not a major factor below 100 m. Blue-green and red algae may extend to this depth (and more) but genera are depth-restricted within this zone. Green algae (utilizing red light) are more restricted with dasycladaceans found today <20 m.

6.2.3.2 Palaeotemperature and palaeoclimate can be assessed in the field by:

1. application of the temperature ranges tolerated by extant taxa (for instance, larger foraminifera, thick walled molluscs and massive scleractinian corals generally indicate warm conditions);
2. morphological features giving an indication of temperature e.g. high proportion of dicotyledonous leaves with serrate margins is typical of humid temperate vegetation, drip tips (tropical, high humidity) leathery leaves and thick cuticles (high temperature evergreens);
3. studying the palaeogeographic maps of the area in question to assess palaeolatitude, and thence an indication of climate;
4. analysis of banding in timber and corals, which is generally associated with seasonality (Fig. 1.2);
5. do not overlook features of sedimentary origin, desiccation cracks, salt pseudomorphs, frosted and rounded grains etc.

6.2.3.3 Type of substrate principally affects trophism (Appendix D), attachment and penetration of benthic organisms. Criteria for distinguishing between soft (loose), firm and hard substrates are given in (§7.7.3). As a generalization, sandy substrates are dominated by filter feeders, and muddy substrates by deposit feeders. In neritic environments trophic diversity (types of feeding) is highest in muddy sand.

When looking at fossils which were part of the original benthos try to determine what the substrate provided for them: (1) simply a location for attachment e.g. an epilithic suspension feeder (oyster); (2) protection e.g. rock-boring clams; (3) a favourable environment for penetration e.g. rapid-burrowing razor shells; and also (4) a source of food e.g. deposit-feeding, burrowing echinoids.

Sequential changes in biota may be related to taphonomic feedback, i.e changes in community structure due to changes in the nature of the substrate associated with dead skeletal material (Fig. 6.10). There are two types:

1. facilitative, e.g. on a soft substrate the accumulation of dead shells provides a substrate for attachment by encrusters (oysters, serpulids) and endobionts (boring bivalves, sponges, worms). The skeletal debris may become the foundation for a buildup, and shell gravels provide refuges and covers;
2. inhibititive, where accumulation of dead shells inhibits bioturbation, infaunal burrowers and deposit feeders by elimination or stunting.

Assuming constant shell input:

1. decrease in sedimentation rate leads to greater shell-packing density, greater abrasion and fragmentation, bioerosion and encrustation;

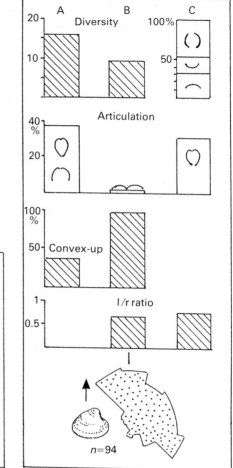

Figure 6.10 (Left) Taphonomic feedback. The accumulation of dead skeletal material may lead to change in substrate: the accumulation of shells of soft-bottom dwellers leads to an increase in the epifauna, and infaunal gravel dwellers (after Kidwell & Jablonski, 1983). (Right) Analyses of three shell beds. Bed A with high diversity, high percentage of articulated shells and relatively low percentage of convex-up orientation indicates parautochthonous storm accumulation. Bed B: analysis of a shell pavement shows low diversity, very low degree of articulation, most valves convex-up and with preferred orientation indicating substantial transport. Bed C with substantial proportion of articulated valves, similar number of right and left valves but poor degree of shell orientation, with respect to bedding, suggesting turbulent wave-dominated processes, rather than unidirectional currents were responsible. The diversity is low (after Fürsich & Heinberg, 1983)

2. increase in sedimentation rate leads to reduced shell packing density, less abrasion, fragmentation, bioerosion and encrustation.

6.2.3.4 Salinity If one is involved with shoreline or lacustrine facies then salinity is of concern. Several chemical methods are available for the determination of ancient salinities, but in the field only faunal or floral evidence can be used. These are, in any case, generally more decisive (*see* Appendix D). Some groups are, with rare exceptions, exclusively stenohaline (articulate brachiopods, echinoids, calcified bryozoans). Certain nekton (cephalopods)

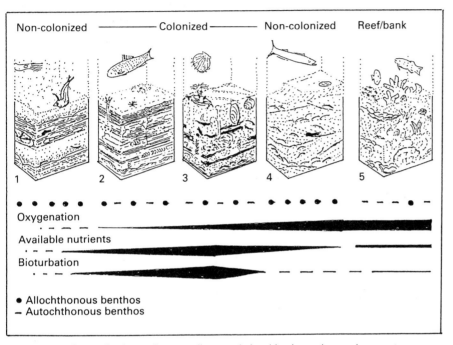

Figure 6.11 Generalized organism – sediment relationships in marine environments. (Applicable to aquatic non-marine environments with modification.) Autochthonous benthos is absent in (1) and (4), sparse in (2). In (3), benthos is mainly parautochthonous, and in (5), dominantly autochthonous. Oxygenation increases left to right. Nutrients available to organisms are at a maximum in environment (3). The sedimentary record is most complete in environment (1). Bioturbation is restricted in (1) and (4) and most typical of (3). Stratification is essentially absent in environment (5), except for local smothering events (storm and volcanic horizons). 1. In environment (1) there is no benthos. The complete sedimentary record (mud and occasional graded silts) shows an entombed cephalopod, and a concretion has enclosed a fish. Another fish decays at the sediment–water interface. 2. In environment (2) there is sufficient oxygen for a sparse benthos. The sedimentation record is fairly complete. 3. Environment (3) is well oxygenated. The water and sediment (typically muddy sand) carries nutrients for a diverse epifauna and infauna of several trophic types. Turbulence is often high enough to erode and redistribute shells. 4. Environment (4) is too unstable to support much life, and the repeated discontinuities record high turbulence. 5. The almost wholly organic environment (5) is 'reefal' and highly oxygenated. On burial, diagenetic changes tend to obscure the growth forms. Typical situations: (1) Euxinic deep basins (Gulf of California); (2) Quiet lagoons, deeper oceans; (3) Continental shelf, Gulf of Mexico; (4) River mouth bars, oolite shoals, offshore bar sands; (5) modern reefs. The diagram also reflects the not uncommon sequence in the fossil record, right to left, of the sedimentary environments found across a marine shelf, nearshore to basin. (Diagram based on Schäfer, 1972: 1. letal-pantostrate facies (non-life – complete sedimentation record); 2. vital-pantostrate facies (with life); 3. vital-lipostrate facies (incomplete sedimentation record); 4. letal-lipostrate facies (non-life and incomplete sedimentation record); 5. astrate facies (absence of stratification) (but most buildups are much more complex – *see* Fig. 3.5)

are marine, but marine fish and aquatic reptiles can invade brackish/freshwater systems. Certain taxa are today restricted to freshwater environments (*Viviparus, Planorbis, Unio,* many ostracodes) or will tolerate minimal salinity. Other genera contain species which are normally freshwater and species which are brackish. Size reduction is typical as a species is traced into brackish water e.g. *Mytilus* in the Baltic. In a lagoonal sequence it may be possible to divide the biota into fresh, brackish and quasi-marine associations (Fig. 8.4).

Consider: (1) that salinity can change rapidly (therefore be on the lookout for rapid stratigraphical change); (2) freshwater forms can be washed into lagoons and the sea, but that; (3) the reverse is less likely; (4) lagoons and gulfs open to the sea will show a range of salinity across the lagoon but salinities will not exceed normal marine salinity; (5) closed or isolated lagoons and lakes can show a range of salinity from zero to halite hypersalinity; (6) the occurrence of freshwater does not necessitate a river supply since rainfall on surrounding swamps can provide a reservoir; (7) a saline wedge may form in an estuary; (8) several freshwater taxa are long-ranging and can be traced back to the Mesozoic; (9) extinct taxa can be used by their association with extant taxa; (10) primary gypsum and halite pseudomorphs may indicate hypersalinity at overlying levels. Ascertain the degree of autochthony present. Shell transport in a lagoon will be due mainly to tidal, storm and river processes. Taxa common in estuarine environments are mainly those tolerant to a wide range of salinity (euryhaline).

6.2.3.5 Oxygenation If there is evidence of hydraulic activity (cross-lamination, cross-stratification etc. or degradational evidence) it is likely that the substrate was more than adequately oxygenated. High organic content (dark colour, bituminous smell), associated with fine lamination (and high lateral persistence) and absence of benthonic organisms, including trace fossils, best indicate anoxia. But note:

1. the oxic–anoxic interface may have been only just above the substrate. Look out for colonizations of local elevations e.g. an upper surface of a large ammonite, upper part of a reptile skeleton (§4.3.4);
2. trace fossils that represent burrowing down from a higher stratigraphic level;
3. the trace fossil *Chondrites* (probably made by a chemosymbiont), by association, seems to have been a useful indicator of minimal oxygenation (Figs 6.12, 6.13), though usually regarded as eurytopic.

Conditions of low oxygenation (dysaerobic) are now being found extensively on continental shelves. The principal causes are water-body stratification associated with a freshwater layer (halocline) or temperature (thermocline), and excess organic matter, generally attributable to plankton blooms and/or high terrestrial input. Correlating degree of oxygenation with the biota and sedimentary structures is more difficult. There seems to be a

Figure 6.12 *Chondrites* – two species displaying colour differences with sediment (near Bairritz, southern France, Upper Cretaceous Flysch). Scale bar = 1 cm

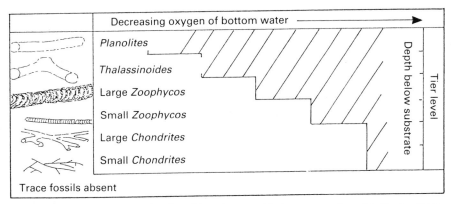

Figure 6.13 Tiering of trace fossils, with main control being oxygenation of bottom waters, in Mesozoic strata (not a general model) (after Bromley and Ekdale, 1984)

gradual increase in burrow diameter with increasing oxygenation. Dysaerobic conditions in the fossil record are associated with large numbers of juveniles but low diversity. In ancient sediments consider the possibility of fluctuating levels of oxygenation or long-term dysaerobic conditions. The type of micro-biota reflects the oxygen levels closely. Occasional winnowing and storm events may lead to shell pavements and shell-filled gutters (§2.6, 7.1). Re-colonization may not have taken place immediately because of hydrogen sulphide release due to the stirring-up of the substrate.

6.2.3.6 Inter-relationship Look for evidence that one organism was influencing another. This is a broad topic, and for clustering see §6.2.1.6, for trophic analysis see §6.2.5 and for sexual dimorphism §6.1. In the field keep a lookout for evidence of predation, bored shells (Fig. 6.14), bite marks (Fig. 6.5), that will aid in determining trophic structure. The other common occurrence is of close association. Most encrusting fossils represent no more than the utilization of a hard substrate (§7.3.3). For evidence of interdependence it is necessary to show that (1) both components were alive together (Fig. 6.15), and (2) the association was beneficial to one or both. In the

Figure 6.14 *Glycymeris* with drill-hole (*Oichnus*), probably made by a naticid gastropod (Pliocene, East Anglia). Scale bar = 1 cm

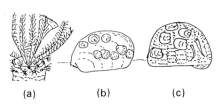

(a)   (b)   (c)

Figure 6.15 Relationships between shells: (a) True symbiosis between serpulid and hydrozoans (reconstruction after Scrutton, 1975) (Lower Cretaceous, Gault Clay, Kent); (b) Bivalve (*Lima*) used as substrate for cemented bivalves (after Seilacher, 1960), (Lower Jurassic). The *Lima* could have been alive or dead at time of colonization; (c) Oysters and serpulids attached to a dead echinoid test (i.e. spines lost). In (b) and (c) the *Lima* and echinoid each acted as a small underwater 'island' for spat settlement

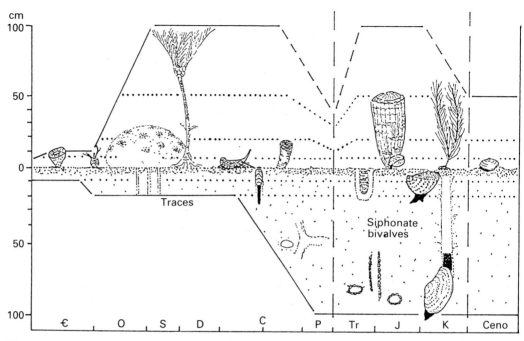

Figure 6.16 A model for tiering of soft substrate suspension feeders, above and below the sediment–water interface through the Phanerozoic in subtidal shallow marine environments. Dotted lines refer to tier divisions (after Ausich and Bottjer 1982, modified in Bottjer and Ausich 1986)

Palaeozoic the shelly biota was dominantly epifaunal so that any time surfaces (bedding surfaces) will, ideally, carry the (fallen) elements of a probably tiered association. In Mesozoic and younger sediments more elements were infaunal, with tiering at various levels above and below the substrate. Figure 6.16 gives an indication of tiering above and below the substrate through the Phanerozoic for suspension feeders. Complications arise as sediment accumulated with overprinting of successive infaunal levels, or from taphonomic feedback (§6.2.3.3). Analysis can be attempted in the same way as is suggested for trace fossils (§7.3.6; Fig. 7.23).

6.2.3.7 Abundance (frequency) is often given as common, rare, very rare etc. Only the author may know what is meant! If such qualifications are used try and specify what is meant in terms of numbers per unit area or volume. For instance, trace fossil X might be described as rare if it was found only once over a quarry with 50 beds accounting for 300 linear metres of accessible soles, or common if present on one in three soles. Shell abundance can be semi-quantified using Schäfer diagrams (Fig. 7.7). For quantified abundance data collect bulk samples or make counts using quadrats (§3.5). With dissociated and fragmented material (the norm for fossils) estimates have to be made e.g.:

brachiopods $\big\}$	n of articulated shells + number of either
bivalves	pedicle or brachial valves (or right and left valves) + fraction of broken shells
gastropods	n of apical portions
cephalopods	n of specimens with a complete whorl
trilobites	n of pygidia (or cranidia) ÷ 10 (number of moults)
crinoids $\big\}$	n is an estimate of number of stem
ophiuroids	ossicles ÷ number (estimate) of stem ossicles per individual
colonial organisms	number of colonies (for branching colonies estimate number of fragments that constituted a whole colony)

6.2.3.8 Diversity Accurate assessment of palaeodiversity depends on what proportion of the original biota is preserved (Fig. 4.3), extractable, determinable and, for the benthos, autochthonous. It is also related to sample size (*see also* §3.5). There are *two* components to diversity: richness (the number of species present), and evenness (the relative abundance of each species; see Appendix C6).

Factors associated with low diversity:

1. High stress, involving any of the physical and chemical limiting factors e.g. salinity decrease from open ocean to enclosed sea (e.g. North Sea to Baltic, Indian Ocean to Persian Gulf), turbulence increase towards exposed shore, or biological factors (e.g. predation pressure).
2. Ecological immaturity, as with opportunistic populations colonizing new substrates. Opportunistic (pioneer) species are much over-represented in the fossil record, and can often be found above event beds, or derived from short-lived environments. Almost only opportunistic ichnospecies are found in deltaic facies though the swamps were occupied by specialists. Most opportunists are characterized by small size and short life span; thus few growth lines on shells. In the field look for thin, relatively widespread (but isochronous) horizons, dominated by, especially, one species of body or trace fossil (e.g. *Skolithos*).

Factors leading to increase or maintenance of diversity:

1. Environmental stability. In a stable environment opportunists are gradually replaced by species with greater capabilities for competitive survival (specialists, equilibrium species of a climax community) and occupying a range of tier levels.
2. Ecological maturity. Today, low latitudes are more stable than high

latitudes, and were less affected by the Pleistocene glaciation. Thus the latitudinal (polar) decrease in diversity is especially marked at present. The deep sea displays high diversity (though low abundance), due mainly to its stability.

3. Where ecological niche availability is high as on a tropical reef, diversity is enhanced. Similarly, in a tropical area with many islands there is greater opportunity for allopatric speciation to occur.

Factors tending to reduce diversity are the opposite of those above: fluctuation in light intensity, salinity (especially in estuarine waters), nutrient supply etc, and reduction in number of niches.

Diversity is highest where ecological disturbance occurs with just sufficient frequency to prevent competitive exclusion. Shallow marine allochthonous assemblages may reflect the time-averaging of a number of populations/communities (intra-habitat) or, a hydraulic concentration from many environments over a vast area (inter-habitat). Skeletal material may be transported hundreds of kilometres by storm action.

In the field:

1. Sample beds for laboratory investigation.
2. Establish trace fossil diversity. For inclusion with body fossil diversity it is necessary to ascertain that the traces belong to the same horizon (see §6.2.4). Note any correspondence between trace and culprit e.g. claw in a burrow. Trace and culprit together is rare e.g. limulid and *Kouphichnium*.
3. Carefully analyse hardgrounds, buildups or any non-collectable strata.
4. Attempt to determine evenness of diversity between assemblages (Appendix C).

6.2.3.9 Dispersion When dealing with an autochthonous assemblage there are three possibilities: a taxon may be randomly or regularly distributed or clumped (Fig. 6.17). For methods of analysis see Appendix C. Clumping can be advantageous for larval settlement or reproduction, but deleterious for growth. Regular distributions minimize competition e.g. in shading, suspension feeding.

Figure 6.17 Random (left), regular (centre) and clumped (right) dispersions

6.2.3.10 Population structure and dynamics What information can be obtained in the field about:

1. The age and size distributions of individuals in a population (of a species): population structure?
2. Rate of population growth or decline: population dynamics?
3. Spatial variation in population structure and dynamics, in similar or different facies? (*See also* dispersion §6.2.3.9)
4. Long-term changes in population structure?

These questions are the same as an ecologist applies. For any satisfactory synecological study an attempt must be made to answer these questions. But the problems facing the geologist are many. To consider a few:

1. It is seldom that the age reached by a fossil shell can be determined. Nor is growth rate constant over life span. A relationship between age and a growth parameter such as shell length will have to be determined and applied. For modern brachiopods log length is approximately equivalent to age.
2. Consider whether the size distribution of a species in an autochthonous shell bed approximates to the original population structure. It may do so with a census population that was 'caught' e.g. below an ash or storm-event bed. Otherwise the fossils represent a graveyard. Only if the recruitment and mortality rates were similar will size/frequency histograms approximate to population structure. There is the exception: moults (instars) of benthic trilobites and ostracodes should approximate to a census population.
3. Assess whether an allochthonous shell bed may be used for analysis of population structure. Do not assume that it cannot (*see* §6.2.3.8).

Generally one is collecting for later analysis. Collect from as narrow (time) intervals as possible. For fossils cemented to bedding surfaces or material which is not feasibly removed, make measurements on individual species, of parameters that are age-related (see text-books), and arrange these in convenient size classes. Ignore compressed and otherwise damaged specimens as far as possible. Analysis is best done having mapped an area or quadrat, so that specimens can be marked as measurement proceeds. Plotting will also generally be carried out later. But it is useful to have in mind the factors that affect a size-frequency histogram:

1. biological: recruitment, including spawning pattern; mortality rate, including seasonal mortality; growth rate; predation rate;
2. environmental: substrate type, temperature etc.;
3. geological: taphonomic processes: sorting, dissolution, fragmentation, collecting failure.

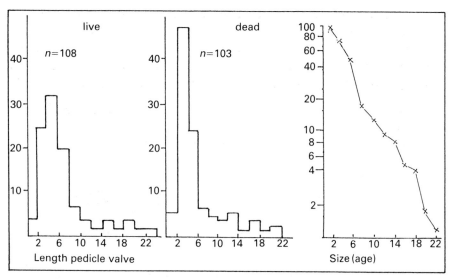

Figure 6.18 (left) Size – frequency histogram of length of pedicle valves of a living population of brachiopods, (centre) size – frequency histogram of pedicle valves from a dead population of the same species, (right) survivorship curve constructed from the live population. No adjustment has been made to correct linear length to age (*see* text)

Example: The positively skewed size-frequency histograms (Fig. 6.18) of some modern brachiopods are similar for live and dead (transported) populations. Mortality and recruitment are more important than variation in growth rate and these are more important than predation and fragmentation in shaping the histogram. But analyses of fossil brachiopods tend to show histograms that are more symmetrical. It may be argued that juvenile mortality is now higher than in the past.

An approximation of a survivorship curve (Fig. 6.18) can be constructed from a size frequency distribution. Attempt to determine the relationship between size and age (above). In plotting, the ordinate should be logarithmic (per cent surviving). At size (age) zero 100 per cent of specimens are 'surviving'. At 'end' of the first size (age) class subtract the numbers of that class from the total number of specimens. These are the 'survivors'. To determine their percentage the calculation for any point on the survivorship curve is:

$$\text{\% population surviving at end of class } x = \frac{\text{total number of population} - \text{sum of specimens in classes between 1 and } x}{\text{total number of population}} \times 100$$

6.2.3.11 Palaeobiogeography   For field observations it is useful to have ˙an appreciation of the principles of animal and plant palaeogeography. What limits a taxon's geographical range? There are just two categories: the chemical, physical and biological limiting factors, and the actual geographical and geomorphological barriers (isthmus, seaway, mountain chain, desert etc.) that restrict attainment of the potential range. Change in the geographical range of a taxon can only come about by breakdown of geographical barriers (by geomorphological, eustatic or tectonic causes) or changes in a taxon's tolerance to the limiting factors (assuming initial dispersal to be quite rapid). Palaeobiogeography is a complex topic. There are four aspects to consider in the field:

1. absence of taxa that might ordinarily be expected to be present (or *vice versa*);
2. differences in composition across tectonically linked regions (§9.3);
3. the apparent migration of taxa across an area because of shift of climatic belts (but this is only applicable to areas of 100s to 1000s km extent);
4. presence (rare) of a taxon that usually occurs in another province e.g. a warm water (tethyan) ammonoid in a cooler (boreal) province. The occurrence may be due to chance floating-in following death.

6.2.4   Biogenic sedimentary structures: trace fossils and bioturbation (For modes of preservation §4.4; stratinomic and sedimentological aspects §7.3)

This section concerns the ecological information that can be derived from trace fossils. Trace fossils represent the life activities of organisms: mainly, animal burrows, trails, trackways, and borings, but also coprolites and faecal pellets. Eggs, skin impressions, calcareous worm tubes are body fossils as no activity was involved. Organisms dragged over, or impinging on the substrate leave tool marks: the organisms were generally dead. With only rare exceptions trace fossils are found exactly where they were made. This is a tremendous advantage over most shells and other body fossils. But it is rare for the culprit, originator, of the trace to be found with the trace, simply because traces represent life activities. To find the culprit and its trace together is an important find and may clinch an otherwise speculative association, but beware of squatters.

There are several points to keep in mind when studying trace fossils:

1. Different organisms, by using similar types of activity, create the same type of structure. For instance, vertical shafts are today made by a variety of worms, arthropods, coelenterates, and this must have been the case in the past.
2. Different structures can be made by the same animal depending on what it is doing (Fig. 6.19).
3. Today, a complex burrow system may be inhabited by several types of animal at the same time.
4. A trace may be taken over by a different type of animal.
5. Pose the question: what factors determine the morphology of any trace fossil? The type of behaviour is the principal factor (Figs 6.19, 6.20). Sediment texture (§6.2.3.3, 7.3.3) obviously influences the form of a

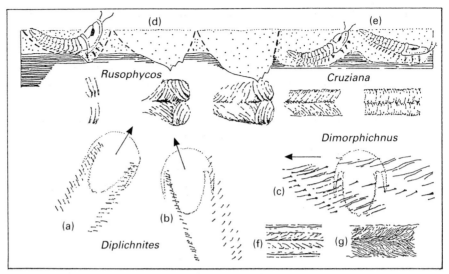

Figure 6.19 Trace fossils made by trilobites to illustrate principles of: 1. Multifunction: differences in trace produced by animal (a) walking straight ahead; (b) slightly obliquely; or, (c) sideways; or making (d) short burrow; or, (e) continuous plough; 2. Polynominal nomenclature; 3. Preservational variation in (d) depending on depth to which animal excavated in mud below sand layer; 4. Variation due to different attitude (e) of animal during burrowing; 5. Stratigraphic value (f) *Cruziana semiplicata* (Upper Cambrian), (g) *C. furcifera* (Lower Ordovician). Ichnogenera are separated by distinct differences in behaviour. Preservational differences are not normally of taxonomic significance (after Seilacher, 1970)

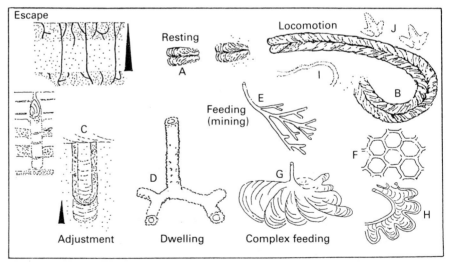

Figure 6.21 *Cruziana furcifera* in typical convex hypichnial preservation (Lower Ordovician, Jordan). Scale bar = 1 cm

Figure 6.20 Ethological categories (behaviour patterns) of trace fossils. (A) *Rusophycus*; (B) *Cruziana* (Fig. 6.21); (C) *Diplocraterion*; (D) *Ophiomorpha* (Fig. 6.22); (E) *Chondrites* (Fig. 6.12); (F) *Paleodictyon* (Fig. 7.11); (G) *Zoophycos* (Fig. 6.25, 6.23); (H) *Phycosiphon*; (I) surface trail; (J) vertebrate track. (C,G,H burrows with spreiten). (Various classically-derived terms are used for the several categories: Cubichnia, temporary resting, Repichnia, locomotion; Fugichnia, escape; Domichnia, domicile, Pascichnia, feeding by grazing systematically over, or just under the substrate; Fodinichnia, feeding by systematically 'mining' the sediment; Agrichnia, feeding by 'farming' or 'trapping' (F). For *Chondrites* see also §6.2.3.5

footprint, the nature of a burrow wall, and hardness, the shape of a boring. Locomotory, escape and feeding burrows are typically unlined.

6. Rate of sedimentation also influences burrow form e.g. escape burrow.
7. The form will also be modified by an animal's growth and adjustment to small fluctuations in many ecological factors.
8. Traces reflect to a greater or lesser extent the morphology of the organisms responsible. This is particularly clear with vertebrate and arthropod traces and, in both, small differences can be utilized stratigraphically (Table 8.4).
9. Apart from some vertebrate and arthropod tracks, trace fossils cannot be classified biotaxonomically (unless associated with the culprit, or are relatively modern). It is generally agreed that the naming of trace fossils should be independent of the naming of the culprit. Figure 6.20 shows how trace fossils can be grouped into a number of ethologic (behavioural) categories. There are problems, e.g. *Ophiomorpha* (Fig. 6.22) represents both a living burrow system and a position for feeding. An ichnogenus defines a characteristic morphology, which can generally be related to a distinctive type of behaviour (cartoon).
10. Some traces have a restricted stratigraphic range, others a long range, and some have shifted habitat (facies) through time.

Figure 6.22 *Ophiomorpha nodosa*, in acetate peel, from loose sand at locality depicted in Fig. 2.5. Scale bar = 1 cm

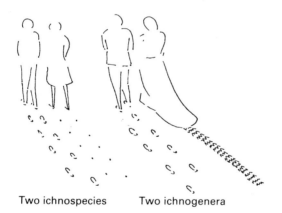

Two ichnospecies   Two ichnogenera

6.2.4.1 Trace fossils and facies analysis (Figs 6.23, 6.24)

The use of trace fossils in environmental interpretation is the one most used and discussed. It is an application quite distinct from those discussed in §7.3. The factors involved are ecological.

As with body fossils, there are two principal stages to analysis: autecological and synecological. The alternative – to pick certain traces and to slot these into described ichnofacies and environmental models – is unscientific. It is worth reminding ourselves that all epifaunal and infaunal organisms have a direct or indirect relationship with a substrate surface. A callianassid shrimp burrowing in loose sand, producing *Ophiomorpha* burrows, can extend down to a depth

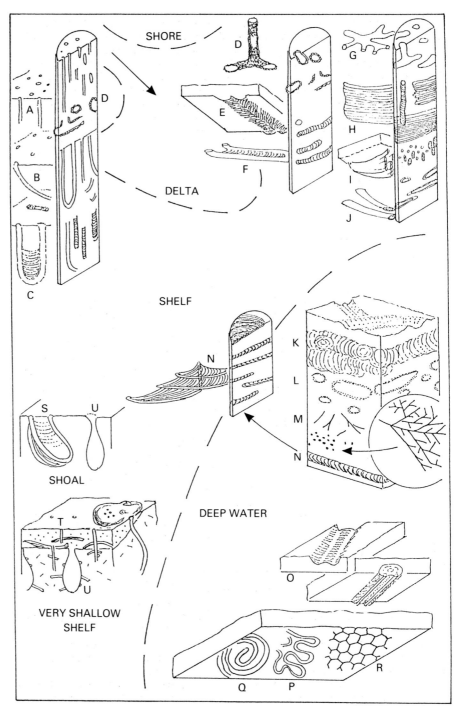

Figure 6.23 Distribution of some trace fossils in marine environments, together with various sections of the ichnogenera as seen in split cores (adapted from Ekdale *et al.*, 1984) (A) *Skolithos*; (B) *Arenicolites*; (C) *Diplocraterion*; (D) *Ophiomorpha* (Fig. 6.22); (E) *Cruziana* (Fig. 6.21); (F) *Rhizocorallium*; (G) *Thalassinoides*; (H) *Teichichnus*; (I) *Phycodes*;

(J) *Terebellina*; (K) *Subphyllochorda* (see O); (L) *Planolites*; (M) *Chondrites*; (N) *Zoophycos* (Fig. 6.25); (O) (above) *Scolicia*, (below) *Subphyllochorda* (exogenic and endogenic echinoid burrows) (also shelf and intertidal); (P) *Cosmoraphe*; (Q) *Spiroraphe*; (R) *Paleodictyon*; (S) *Glossifungites* (=*Rhizocorallium jenense*); (T) *Trypanites*; (U) *Gastrochaenolites*. Trace fossils are commonly grouped into ichnofacies: *Skolithos* ichnofacies (A–C), *Cruziana* ichnofacies (E–J), *Zoophycos* ichnofacies (N), *Nereites* ichnofacies (O–R), *Trypanites* ichnofacies (T, U), *Glossifungites* ichnofacies (S, U). (D), (L) and (M) are generally referred to as facies-crossing traces. (block diagram of tiering after Wetzel, 1984). Note that in split cores it is often not possible to assign a burrow to a definite ichnotaxon. Use a temporary, simple descriptive term, with dimensions. Use the diagram (Appendix C) to determine the orientation of cylindrical burrows running obliquely to the rock face or split core. Scan breaks in cores for bedding-parallel traces. Break cores at the base of a bioturbated zone or event bed to locate discrete traces

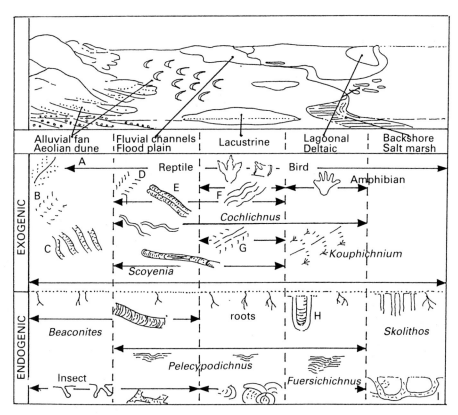

Figure 6.24   Distribution of some trace fossils in non-marine environments (after J E Pollard, 1980 ©): (A) *Paleohelcura* (scorpion tracks); (B) *Mesichnium* (insect tracks); (C) *Entradichnus* (insect tracks and burrows); (D) *Acripes* (crustacean tracks); (E) *Isopodichnus* (phyllopod or notostracan burrow); (F) *Undichnus* (fish traces); (G) *Siskemia* (arthropod track); (H) *Diplocraterion*; (I) *Psilonichnus* (crab burrows). *Beaconites* (probably vertebrate burrow); *Cochlichnus* (worm trail, burrow); *Kouphichnium* (limulid (xiphosurid) tracks); *Scoyenia* (insect trail or backfilled burrow)

of a couple of metres or more and then open up into galleries. The sands adjacent to the galleries may belong to a variety of depositional environments. But, the substrate surface that was colonized above is likely to have been shallow marine, and it may well have been penecontemporaneously eroded and now missing.

Pose the question: what sedimentary structures and trace fossils might be expected in the environment being postulated. Some expected features may be absent. Why?

A few types of trace fossils tend to be very common and widespread. For instance, *Chondrites* (Fig. 6.12) is often the first trace to appear above an apparently anaerobic sequence of muddy sediments, suggesting that the animal responsible had a tolerance to low oxygen. The common *Ophiomorpha* obviously owes its frequency in part to the high preservation potential of its deep burrows, but a modern culprit, *Callianassa major*, is associated particularly with unstable sand substrates. In contrast, the complex networks *Paleodictyon*, or the complex spirals *Zoophycos* (Fig. 6.25) suggest that their construction took a substantial period of time. Their occurrence is always in relatively stable environments. Ecological factors deducible from various behavioural types are shown in Table 6.7. Depth is not, in itself, an important control. Do not be alarmed if 'shallow-water' traces occasionally occur in what is interpreted as a deep-water facies (especially if more proximal), and *vice versa*, e.g. *Zoophycos* in lagoonal sediments. But, if shallow-water traces are dominant, have another think. When dealing with event beds, especially turbidites, take care to distinguish between the pre- and post-event suites (§7.3.1).

A general scheme for the *quantification* of the extent of bioturbation in a sediment is impossible, for the same reason that it is meaningless, in terms of estimating abundance, to record the number of shells in a given volume of

Figure 6.25 *Zoophycos brianteus.* [Helminthoid Flysch, Middle Eocene, near Palma, Italy (with spreiten)]. Scale bar = 1 cm

Table 6.7 Ecological factors deducible from trace fossil ethology

Resting	1.	nature of substrate
	2.	level of turbulence
	3.	environmental stability
	4.	availability of food from water or sediment
	5.	oxygenation
	6.	direction of palaeocurrent: rheotaxis
Escape	1, 5.	
	7.	environmental instability
Adjustment	1.	
	8.	periodic instability of sedimentation rate (aggradation or degradation)
Deposit feeding Complex feeding, grazing	1, 5. 8.	food resources, even/uneven distribution
Locomotion Dwelling	1, 5.	stability of all environmental factors

sediment from different facies. Trace fossils also vary in size. But, it can be useful, for a given facies, to record the extent of bioturbation, since this will be a measure of stability as well as an indicator of other ecological factors. Comparison (flash) cards can be constructed for field use (Fig. 7.8). (For ichnofabric: see §7.3.6.)

6.2.4.2 Description of trace fossils is generally necessary in the field since many are too large or too difficult to extract (*see also* §7.3.6, Fig. C 3).

1. Note location.
2. Ascertain the stratigraphy and facies.
3. Try to understand the mode of preservation of each trace (4.4).
4. Begin each description with overall form, dimensions, relationship and attitude to bedding (parallel, oblique, vertical); then go to details of form e.g. branching, sculpture, termination type, nature of burrow fill and burrow wall, form of spreite.
5. Attempt to establish the relative timing of formation of each type of trace (ordering) and the tiering. These are discussed further in §7.3.6, and both can be difficult and time-consuming.
6. Sketch and photograph.
7. When collecting take care to sample any variation that may be due to behavioural differences. Where applicable, always collect counterpart material for stratinomical analysis.

In describing vertebrate trackways map the distribution of prints as accurately as possible. Also, make the measurements indicated on Fig. 6.26.

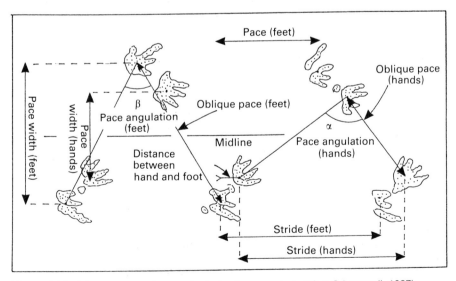

Figure 6.26 Measurements for vertebrate trackway analysis (after G Leonardi, 1987)

6.2.5   Palaeocommunities and palaeoenergetics

At quite an early stage in an investigation of the palaeoecology of a fossiliferous site it is useful to attempt a pictorial reconstruction of the site as it was at
the time of formation. This focuses one's mind wonderfully on asking the pertinent questions. Later a pictorial reconstruction of the palaeocommunity can
be made, which can be fully justified from the analyses.

Many substrate reconstructions have been published, but few have been
based on a quantitative assessment of the sedimentology and palaeontology.
Many are no more than inspired guesses. Of course, any reconstruction will
depend to a considerable extent on how community is defined. A useful and
practical definition for palaeontologists is that a palaeocommunity is an association of organisms that previously lived together. This places emphasis on
assessment of autochthony. A definition that will involve more sophisticated
analysis is that a palaeocommunity is an association of organisms which, when
alive, formed an interdependent association. This will involve statistical tests
and determinations of former biomass and energy flow. Further discussion
is beyond the scope of this section, but it is useful to consider the major problems confronting analysis of fossil material: (1) that energy flow is only determinable on the basis of an individual's life span, not on a daily/annual basis
(as in modern ecology), (2) the relationship between shell size (all that
remains) and the former biomass is not readily determined or estimated, (3)
no adequate allowance can be made for the soft-bodied biota, (4) taphonomic
loss was probably greater than it appears to be.

As work progresses on modern substrates, especially on shelf benthos,
we are gaining a better appreciation of how modern communities can be
defined and how they might become fossilized. Particularly relevant to
ancient sequences is the appreciation of short period fluctuations, and long
term changes in population structure. One might expect these to be well
represented in the fossil record; in the development of fossil reefs, changing
successions on soft substrates and so on. But naturally occurring 'autogenic'
successions, where the succession is biologically controlled, but not necessarily deterministically so, are not easy to prove and, in most cases, physical and
chemical processes have intervened, thus leading to community replacement.

A simple method of analysing fossiliferous sediments is to construct a
trophic–substrate–mobility chart (Fig. 6.27) followed by a sketch of the substrate reconstruction. An indication of relative abundance can be included
together with an indication of the tiering above and below the substrate of
body and trace fossils. Feeding habits and habitats can also be presented on
ternary diagrams (Fig. 1.2). Plotting should be done as species abundance
rather than species presence, because some species may be represented by few
specimens. Use of such diagrams has been criticized because, when applied to
modern communities, analysis of the dead (shelly) fauna (analogous to the
fossil material) is distorted when compared with that of the live fauna. Similarly suspect is the use of trophic nucleus (number of species that account for

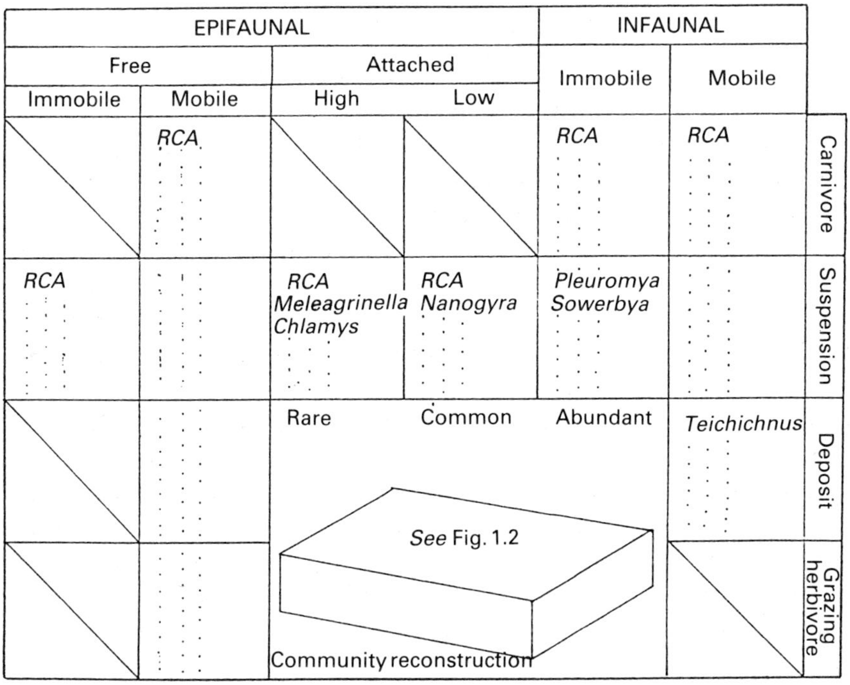

EPIFAUNAL				INFAUNAL		
Free		Attached		Immobile	Mobile	
Immobile	Mobile	High	Low			
	RCA			RCA	RCA	Carnivore
RCA		RCA Meleagrinella Chlamys	RCA Nanogyra	Pleuromya Sowerbya		Suspension
		Rare	Common	Abundant	Teichichnus	Deposit
		See Fig. 1.2				Grazing herbivore
		Community reconstruction				

Figure 6.27 Trophic–substratum–mobility chart, and community reconstruction, with an example from the Jurassic depicted in Fig. 1.2

80 per cent; Fig. 1.2), but both may be useful in facies analysis. If the data are sufficient, more sophisticated statistical analysis can be applied to discriminate the palaeocommunities.

6.3 Fossil-ores (Fossil-Lagerstätten; Fig. 6.28)

There are a number of fossiliferous localities known to just about every geologist: the Solnhofen Limestone with *Archaeopteryx*, the Burgess Shale with many soft-bodied animals and trilobites with preserved appendages, or the Rhynie Chert with the exquisite preservation of early vascular plants. There are many others. Such sites are often referred to as Fossil-Lagerstätten or sites of extraordinary preservation. Many result however from the coincidence of normal taphonomic processes with normal sedimentological situations and biological evolution. A careful sedimentological analysis of the fossiliferous site may reveal the mechanisms that gave rise to the preservation.

Using the term fossil-ore reminds us that the mining geologist locates ore bodies on a scientific basis, by mapping, prospecting and predicting where to recommend mining with a defined degree of confidence. There are four main categories of primary fossil-ore.

1. Many autochthonous buildups (reefs; §3.6).

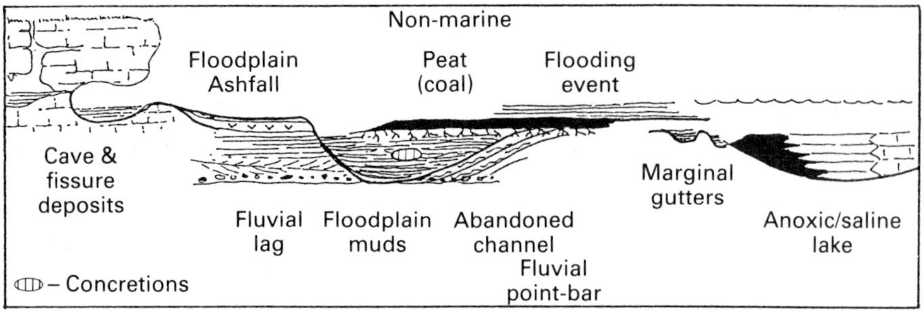

Figure 6.28   Environmental settings of some primary fossil-ores (adapted from Seilacher & Westphal, in Whittington & Conway Morris, 1985); *see also* Fig. 1.1

2. Occurrences characterized by an abundance of fossil remains due to hydraulic processes, sorting or winnowing material (e.g. winnowed bone concentrates, richly fossiliferous cross-stratified gravels, or fissure fills representing hydraulic traps (sieves).

3. Occurrences characterized by articulated skeletons (especially of vertebrates and echinoderms). The most important factor is absence of bioturbation and reworking, hence seek deposits representing rapid covering (§7.3.1), rapid burial by submergence in muddy 'liquid' sediment (deforming any lamination), incorporation into a high density flow (wackestone) (Burgess Shale situation), or amber, or deposits indicating an anaerobic substrate. Most large marine vertebrates are found in mudrocks. Expect differences in the ability of live animals to have escaped from such situations (low for echinoderms and trilobites), and variation in the way dead material may have responded.

4. Occurrences characterized by the presence of soft tissue as well as completeness (*see* §4.3.1). These are more common than is generally thought.

A crinoid complete with stems and arms attached to the calyx, or a trilobite virtually complete, or a pocket of well-preserved brachiopods, is a minor fossil-ore. Pose the appropriate question: might there be a brittle star on the sole of this storm bed? Or, might there be insect remains associated with all these plant fossils?

Secondary 'fossil-ores' are where fossils are reworked from any of the major types of fossiliferous sediment (Chapter 1) and concentrated by modern processes. An example is the wave-concentrate along Kent (southern England) beaches of pyritised twigs, seeds and fruits derived from the cliffs of London Clay (Eocene), a type 5 fossil deposit. Another is the strip-mine dumps at Mazon Creek (Illinois) with fossiliferous nodules from a type 10 fossil deposit.

Looking at fossiliferous sites in this way helps to give a better appreciation of the fossil record. But with many fossil-ores there is a price to be paid. Most

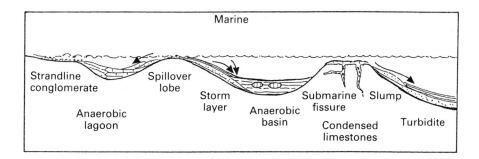

have a strongly biased biota, e.g. general absence of benthos at Solnhofen, loss of aragonitic fossils in the Faringdon Sponge Gravels. The rare occurrences of the preservation of a high proportion of the original assemblage, as has occurred in the Silesian of Mazon Creek, in the Burgess Shale (Middle Cambrian) and in the Eocene of Messel are all the more important.

References

AUSICH W I, BOTTJER D J 1982 Tiering in suspension-feeding communities on soft substrata throughout the Phanerozoic. *Science* **216**: 173–4

BOSENCE D W J, ROWLANDS R J, QUINE M L 1985 Sedimentology and budget of a Recent carbonate mound, Florida Keys. *Sedimentology* **32**: 317–43

BOTTJER D J, AUSICH W I 1986 Phanerozoic development of tiering in soft substrate suspension-feeding communities. *Paleobiology* **12**: 400–20

BROMLEY R G, EKDALE A A 1984 *Chondrites*: a trace fossil indicator of anoxia in sediments. *Science* **224**: 872–4

CRAME J A 1981 Ecological stratification in the Pleistocene coral reefs of the Kenya coast. *Palaeontology* **24**: 609–46

FÜRSICH F T, HEINBERG C 1983 Sedimentology, biostratinomy and palaeoecology of an Upper Jurassic offshore sand bar complex (Greenland). *Bulletin of the Geological Society of Denmark* **32**: 67–95

KERSHAW S, RIDING R 1978 Parameterization of stromatoporoid shape. *Lethaia* **11**: 233–42

KIDWELL S M, JABLONSKI D 1983 Taphonomic feedback: ecological consequences of shell accumulation. *In* Tevesz M J S, McCall P L (eds) *Biotic Interactions in Recent and Fossil Benthic Communities* Plenum p. 195–248

LEONARDI G ed 1987 *Glossary and Manual of Tetrapod Footprint Paleoichnology.* Brasilia, 117 pp

NOBLE J P A, LOGAN A 1981 Size-frequency distributions and taphonomy of brachiopods: a recent model. *Palaeogeography, Palaeoclimatology, Palaeoecology* **36**: 87–105

PAUL C R C, BOCKERLIE J F 1983 Evolution and functional morphology of the cystoid *Sphaeronites* in Britain and Scandinavia. *Palaeontology* **26**: 687–734

POWELL E N, STANTON R J 1985 Estimating biomass and energy flow of molluscs in palaeo-communities. *Palaeontology* **28**: 1–34

SCHOLLE P A, BEDOUT D G, MOORE C H (eds) 1983 *Carbonate Depositional Environments.* American Association of Petroleum Geologists Memoir **33**: 708 pp

SCRUTTON C T 1975 Hydroid-serpulid symbiosis in the Mesozoic and Tertiary. *Palaeontology* **18**: 225–74

SEILACHER A 1960 Epizoans as a key to ammonoid ecology. *Journal of Paleontology* **34**: 189–93

SEILACHER A 1970 *Cruziana* stratigraphy of 'non-fossiliferous' Palaeozoic sandstones. *In* Crimes T P, Harper J C (eds) *Trace Fossils* Seel House Press, p. 447–76

WETZEL A 1984 Bioturbation in deep-sea fine-grained sediments: influence of sediment texture, turbidite frequency and rates of environmental change. *In* Stow D A V, Piper D J W (eds) *Fine-grained Sediments: Deep Water Processes and Facies* Blackwell, p. 595–608

General text on palaeontology

RAUP D M, STANLEY S M 1978 *Principles of Palaeontology* 2nd edn, Freeman, 481 pp

Texts on invertebrate palaeontology

BOARDMAN R S, CHEETHAM A H, ROWELL M (eds) 1987 *Fossil Invertebates* Blackwell, 731 pp

CLARKSON E N K 1986 *Invertebrate Palaeontology and Evolution* 2nd edn, George Allen & Unwin, 382 pp

LEHMANN U 1976 *The Ammonites: Their Life and Their World* Cambridge University Press, 246 pp

SMITH A 1984 *Echinoid Palaeobiology* George Allen & Unwin, 190 pp

Texts on palaeobotany

STEWART W N 1983 *Paleobotany and the Evolution of Plants* Cambridge, 405 pp

TAYLOR T N 1981 *Paleobotany: An Introduction to Fossil Plant Biology* McGraw Hill, 589 pp

THOMAS B A, SPICER R A 1986 *The Evolution and Palaeobiology of Land Plants* Croom Helm, 309 pp

WRAY J L 1977 *Calcareous Algae* Elsevier, Amsterdam

Texts on micropalaeontology

BIGNOT G 1985 *Elements of Micropaleontology* Graham & Trotman, 217 pp

BRASIER M D 1980 *Microfossils* George Allen & Unwin, 193 pp

Texts on vertebrate palaeontology

CARROLL R 1987 *Vertebrate Palaeontology and Evolution* Freeman, 698 pp

BENTON M J 1990 *Vertebrate Palaeontology: Biology and Evolution* Unwin Hyman, 377 pp

Texts on trace fossils

CURRAN H A (ed) 1985 *Biogenic Structures: Their Use in Interpreting Depositional Environments* SEPM, Special Publication No. **35** 347 pp (A collection of 20 papers)

BROMLEY R G 1990 *Trace Fossils: Biology and Taphonomy* Unwin Hyman, 280 pp

EKDALE A A, BROMLEY R G, PEMBERTON S G 1984 *Ichnology* SEPM, Short Course Notes **15** 317 pp

FREY R W (ed) 1975 *The Study of Trace Fossils* Springer, 562 pp

Texts on paleosols

REINHARDT R, SIGLEO W R 1988 Paleosols and weathering through geologic time: principles and applications. *Special Paper of the Geological Society of America* **261** 181 pp

WRIGHT V P (ed) 1986 *Paleosols: Their Recognition and Interpretation* Blackwells, 315 pp

Texts on ecology and palaeoecology

AGER D V 1963 *Paleoecology* McGraw Hill, 371 pp

BARNES R S K, HUGHES R N 1982 *An Introduction to Marine Ecology* Blackwell, 339 pp

DODD J R, STANTON R J 1990 *Paleoecology, Concepts and Applications* 2nd edn, Wiley, 520 pp

GRAY J (ed) 1988 *Palaeolimnology, Aspects of Freshwater Palaeoecology and Biogeography* Elsevier, 678 pp (Book edition with index of *Palaeogeography, Palaeoclimatology, Palaeoecology* **62**: 1–623)

GRAY J S 1981 *The Ecology of Marine Sediments* Cambridge University Press, Cambridge, 185 pp

ODUM E P 1983 *Basic Ecology* Saunders, 613 pp

Palaios: 1988 Volume **3** (4) collection of papers devoted to marine hard substrate communities

Palaios 1988 **3** (2) Ancient reef ecosystems theme issue (12 papers on modern approach to topic)

SCHÄFER W 1972 *Ecology and Palaeoecology of Marine Environments* Oliver & Boyd, 568 pp

SCHOPF T J M 1980 *Paleoceanography* Harvard University Press, 341 pp

TAVESZ M J S, MCCALL P L (eds) 1983 *Biotic Interactions in Recent and Fossil Benthic Communities* Plenum Press, 837 pp (A useful 'state-of-the-art' on geologists' thinking on this topic)

WARNER G F 1984 *Diving and Marine Biology. The Ecology of the Sublittoral* Cambridge Studies in Modern Biology **3**, Cambridge University Press, 210 pp

Texts and References to Fossil-Ores (Fossil-Lagerstätten)

ALLISON P A 1988 Konservat-Lagerstätten: cause and classification. *Paleobiology* **14**: 331–44

GLAESSNER M 1984 *The Dawn of Animal Life* Cambridge University Press, Cambridge, 244 pp

GRANDE L 1984 Paleontology of the Green River Formation with a review of the fish fauna. *Geological Survey of Wyoming, Bulletin* **63**, 333 pp

NITECKI M H (ed) 1979 *Mazon Creek Fossils* Academic Press, 581 pp

WHITTINGTON H B 1985 *The Burgess Shale* Yale University Press, 151 pp

WHITTINGTON H B, CONWAY MORRIS S (eds) 1985 Extraordinary fossil biotas: their ecology and evolutionary significance. *Philosophical Transactions of the Royal Society of London* 192 pp (A collection of papers covering fossil ores of types 3 and 4)

7

Fossils for the sedimentologist

'Without fossils, sedimentology is lifeless and timeless' John Pollard, speaking at the first Palaeontological Association Review Seminar, Reading, Autumn 1980

An account of sedimentary environments and facies is beyond the scope of this volume and reference must be made to the texts listed at the end of this chapter. Figure 7.1 is included to give a brief summary of the main sedimentary environments.

This chapter emphasises the role body and trace fossils play in understanding the hydraulic regime that operated when a sediment was laid down. It also examines evidence for sedimentation rate and for the consistency of the substrate.

In the first section we will examine allochthonous shell accumulations where fossils have been carried by hydraulic processes to their final position in the sediment.

7.1   Allochthonous skeletal accumulations

The skeletons of plants and animals can form accumulations in life position (as in the case of coral reefs or other buildups) or can be accumulated by non-biological sedimentary processes. We are concerned here with the latter type, but remember that reefs may often show areas of allochthonous accumulation as a result of factors such as storm damage and winnowing.

When a skeletal accumulation is encountered in the field, firstly determine its overall geometrical form, then note the nature of its bounding surfaces (it may, for example, have an erosional base and a gradational top). Note also orientation data provided by the individual skeletal remains.

The following list categorises the features and the significance of the main types of allochthonous fossil accumulation.

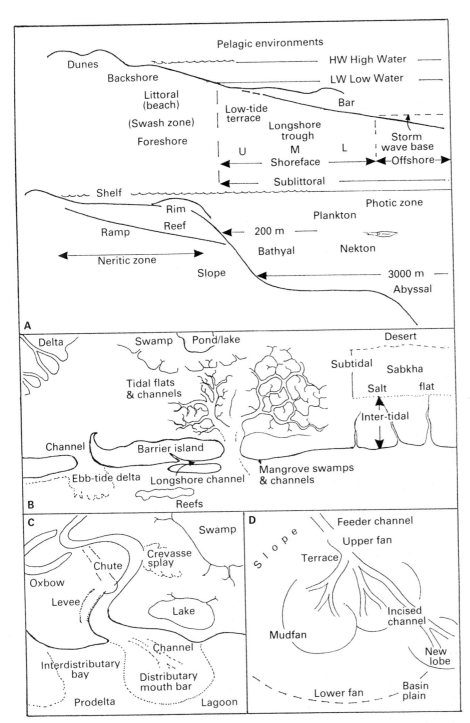

Figure 7.1 (A) Horizontal and vertical zonation of sea and ocean; (B) Shoreline environments with temperate and tropical (wet and arid), intertidal geomorphologies; (C) Coal Measure environments (after Scott, 1979); (D) Turbidite environments

Table 7.1 Procedure for describing allochthonous skeletal accumulations

1. Identify the type of accumulation and its geometry
2. Determine the nature of the fabric and preferred orientation
3. Identify the nature of the matrix
4. Consider the taphonomic history, faunal composition and diversity

Table 7.2 A classification of allochthonous skeletal accumulations (adapted from Kidwell, Fürsich & Aigner, 1986). Bar scale 10 cm

1. Local (cm–m)

    (a) Seen on bedding surfaces
(i) stringers	(iii) flute cast fills
(ii) pavements	(iv) current shadows

    (b) Seen generally normal to bedding
(i) winnowed layers	(iv) desiccation structures
(ii) gutter fills	(v) solution fissures
(iii) burrow fills	(vi) tectonic fissures

2. Extensive (m–km)
(i) lags	(iii) storm beds
(ii) relict accumulations	(iv) turbidites

3. Other types (see text)

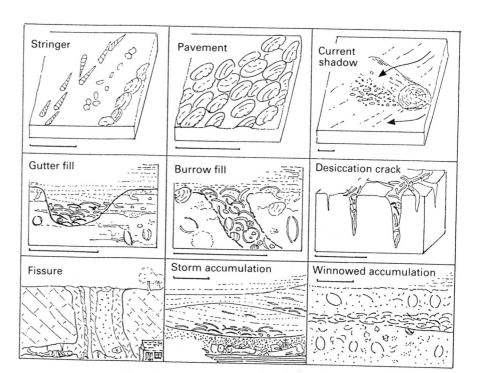

Figure 7.2 Types of allochthonous skeletal accumulation (After Kidwell, Fürsich & Aigner, 1986). Bar scale = 10 cm

1. *Local scale* (cm–m) The fossils are often locally derived, though accumulations associated with solution and tectonic fissures are likely to show complex derivations.
   (a) Generally seen on bedding surfaces:
       Stringers (Table 7.3) Linear arrangements of shells, including ostracode valves, coprolites and faecal pellets are quite common in muddy sediments. They can be useful in giving evidence of higher energy events in sediment more or less devoid of coarse siliciclastic grains. Two types can be distinguished in the field. Type 1 (more common) is current-formed, akin to current lineation, Type 2 is wave-formed, akin to starved (isolate) ripples. On modern beaches with wave ripple, fine and light particles tend to accumulate in the ripple troughs.

Table 7.3 Distinctions between wave- and current-formed linear shell accumulations (after Futterer, 1982)

	Waves	Currents
Distance between shell bands	uniform, as ripple wavelength	variable
Frequency of shell bands	always numerous	often isolated, rarely numerous
Shell attitude within bands	long axes parallel to bands	imbrication in one direction
Deposition of small biogenic particles	on each side of band	only on lee side
Direction of gutter casts	transverse to bands	parallel to bands

Pavements There is probably a gradation from stringers to pavements formed by a layer of shells one, or no more than two, valves thick. Pavements (plasters) are also common in mud rocks. The preferred orientation and possible imbrication of the elements will be a clue to transport direction. Check that the size distribution and shell attitude does not oppose interpretation as a hydraulic concentration (§6.2.1).
       Flute casts may be filled with shells.
       Current shadows Fine skeletal material or silt and fine sand may accumulate in the lee of a larger shell to form a current shadow.
   (b) Generally seen normal to bedding. (Bedding expression must also be studied.)
       Winnowed, or washed, accumulations describe the gradual removal of finer particles by current or wave action. The essential question to ask is whether the sole is sharp or gradational. Winnowing will leave a gradational sole. But in practice, by

following the level laterally, a site can be often found where actual erosion occurred and the shells lie on a sharply truncated surface.

Gutter fills (casts), defining storm track (possibly depositional slope), are generally cm width (rarely dm) and cm depth. Soles are generally rounded in cross-section. Note whether the gutters are straight or sinuous, and whether groups are present. Are there minor groovings on the sides? Is the top rippled? If so, by what type of ripple? Is there more than one episode of cut and fill? Transport direction will probably be apparent only by slabbing. Could the fill material have been derived from the adjacent facies?

Burrow fills arise in two ways: either shelly material is washed into the burrow, or material is selected out by the burrower and biogenically pressed into the backfill. Note that some burrows have a purpose-built wall of shell material e.g. tube worms *Lanice* and *Diopatra*.

Desiccation structures may become shell filled. Check that desiccation was the process forming the crack, by determining the configuration of the crack and its plan. (Syneresis cracks made under water are generally much narrower and incomplete, but may also fill with shells, e.g. ostracodes and shell fragments.)

Solution fissures mostly result from subaerial, karstic processes. When submergence has followed, they may not be easy to distinguish from sedimentary (Neptunian) dykes of tectonic origin. Search for features indicative of a subaerial origin. Terrestrial caves and fissures may yield richly and be extensive.

Fossils found in tectonic fissures are always of interest, because they often represent events for which no stratiform record is preserved.

2. *More extensive sheets or lenticular skeletal accumulations* (cm–m thickness, metre–kilometre extent). Modern storm beds seem to comprise mainly locally derived material, but anticipate the possibility of recycling and distant derivation.

Lags   The most common type of lag deposit is the coarse material at the base of a hydraulically-transported bed of sand. The basal conglomerate to a transgression surface may also be included. There will always be an erosion surface below the lag, and this distinguishes it from a winnowed or washed accumulation. Imbrication may be evident, and cross-lamination may be discernible, depending on the shapes of the materials. Reworked material will generally have been through early stratinomic and diagenetic processes, so that, for instance, the pores of bony or echinoderm elements become infilled with mineral matter or clay grade sediment, thereby much increasing the specific gravity: fresh bone 2.0, fossil bone 2.6–2.9. Such material is referred to as prefossilized. 'Beach' lags generally show high lateral continuity, whereas lags formed during lateral migration (e.g. meandering streams) may show frequent discontinuities with rapid lateral change in composition.

Relict accumulations of skeletal debris are widespread today on continental shelves. Following the rapid eustatic rise of the last 10,000 years, and the changed conditions of light, food supply and low net sedimentation, the dead shells and skeletons remain on the sea floor, or accumulate over a long period. Ancient accumulations of relict material may be difficult to prove, but there is likely to be evidence of prolonged exposure on the sea floor: encrustation, extensive borings (post-Silurian) and carbonate dissolution.

Storm beds (§3.3) being a feature of shelf sedimentation, often carry abundant shell material (Figs 7.9, 7.13). If this is absent, then pose the question, why? (Climate? High rate of sedimentation?). More proximal sequences tend to have beds that are thicker, represent several events, and carry shells which are allochthonous. Distal storm beds are thinner, represent only a single event and carry parautochthonous shells (Fig. 7.3).

Siliciclastic turbidites often carry a sparse biota of shells, plant debris etc. The shells may give an indication of the nature and depth of the source of the turbidity current e.g. beach or seamount (*see* Figs 1.1, 8.8).

3. *Other types of shell accumulations* include extensive, thick, and complex accumulations, pelagic accumulations (§1.3–7,8), clusters (§6.2.1).

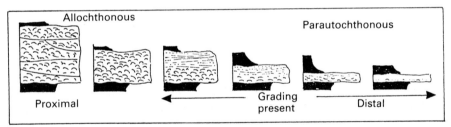

Figure 7.3 Proximal to distal sequence in a shell bed (After Fürsich & Oschmann 1986).

7.2   Other parameters

Having established the type of accumulation and its overall geometrical form, the next task is to analyse the fabric, orientation and preferred orientation of elements, the role of matrix, and to get some appreciation of the diagenesis.

7.2.1   Fabric

To determine *fabric* (Fig. 7.4), it is necessary to understand the three-dimensional organization of the material: orientation, packing and sorting. Is the sediment matrix or bioclast supported? The nature of the exposure may be insufficient to determine all of these, and it may be necessary to collect oriented material for later slabbing. First look for the obvious indications that give a clue

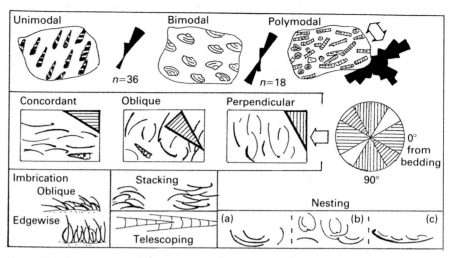

Figure 7.4 Diagram to illustrate terminology for shell orientation on bedding surfaces and in cross-section. (After Kidwell, Fürsich & Aigner 1986) Unimodal, unidirectional current; bimodal, wave oscillation; concordant/convex down, settling from suspension; perpendicular/ edgewise, oscillatory flow; oblique imbrication, up-current dip; stacking/nesting, storm reworking

to process: indications of cross-lamination, hummocky cross-stratification associated sole structures, imbrication, lineation, stringers, attitude of convex shells, groupings of shells.

7.2.2   Preferred orientation

Fossils show *preferred orientation* for two causes: (1) they are in life position and took up their attitude under the influence of light, gravity or nutrient-supplying currents, or (2) they were oriented after death by wave, or current action, or biogenic agencies. What is generally required is an indication of hydraulic strength and direction or trend. Preferred orientation is best measured on bedding surfaces. Measure the angle to the dip direction with a protractor, or from a non-digital watch laid parallel to the dip. Measure separately *trends* (0–180°) e.g. trilobite axes, brachiopod hinge lines, and *directions* (0–360°) e.g. gastropod apices, brachiopod commissures. For plotting rose diagrams use 10°,15°, 20°, or 30° intervals depending on number of readings. With dips greater than 25°, correct the vector mean for tectonic tilt using a stereonet. Note that current or wave-motion needs to act over a period of time to induce shells into position of least resistance. A firm bed is also required, because on a soft/loose substrate, elements tend to become buried.

The orientation of shells etc. in currents is reasonably predictable from their shape. Figure 7.5 shows various types of shells and their orientation under flume conditions. Flattish shells tend to form imbricate arrangements readily. Tall spiral gastropods rotate about the apertural end, with the apex tending to

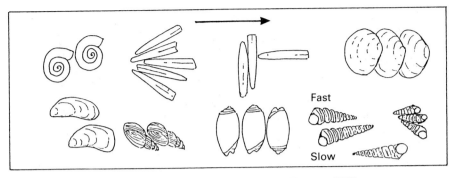

Figure 7.5 Orientations of some shells in a current. (After Futterer 1977)

Table 7.4 Distinctions between wave and current generated edgewise shell accumulations (after Futterer, 1982)

	Waves	Currents
Direction of long axes	different orientations	± uniform, at right angles to current
Inclination	various directions	constant, generally upcurrent
Accumulations	stacks in various directions	imbrication in one direction (down-current)

point downstream, except under slow flow. Surprisingly, even coiled cephalopods may become oriented with the aperture more or less downcurrent, and heteromorphs even more so. Strong wave action (seen on modern beaches) leads to edgewise orientation (Fig 7.4, Table 7.4).

Small shells are not uncommonly concave-up in association with small scale cross-lamination. Elongate particles (wood, crinoid stem lengths) may roll or align parallel to current. A T-arrangement (belemnites, crinoid stems) is a good indicator, the cross-bar being upcurrent. More complex arrangements of crinoid stems have been observed in the field and reproduced in the flume. Belemnites may show complex patterns, because they tend to roll along a curved path. In muddy sediments compaction may make orientation measurement difficult.

Animals with flexible or protruding parts, such as starfish and brittlestars, have been found below a smothering sediment with the arms drawn out – a rare occurrence. In contrast, the more common tangled mass of crinoids around a log of wood in Jurassic mudrock indicates low energy.

Rapid sedimentation is generally associated with a low degree of preferred orientation, poor sorting, high frequency of convex-down valves and, possibly, high numbers of articulated shells. Note that bioturbation may also lead to the first three of these attributes.

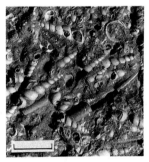

Figure 7.6 Shell pavement with gastropods and bivalve shells. Oligocene, Isle of Wight. Scale bar = 1 cm

7.2.3   Matrix

Matrix is hydraulically significant only if it can be shown to be primary, and not due to infiltration (giving non-uniform distribution and umbrella-effects), or to compacted pellets (identity preserved in shelter cavities (Fig. 4.19)).

Collect material for laboratory slabbing. Distinguish between clast-supported (packstone, grainstone) and matrix-supported (wackestone) fabrics.

7.2.4   Diagenesis

Diagenesis may modify an assemblage considerably. In coarse, loose sediment, aragonitic shells may be completely lost, or there may remain only rare indications of their original presence. In contrast, the number of (originally) aragonitic shells may be relatively enhanced in a shell bed where early cementation has taken place, while only rare specimens remain in life position in an underlying leached siltstone.

7.2.5   Quantitative description

The *quantitative description* of a skeletal accumulation does not generally in itself have a great deal of hydraulic significance. The parameters of size, shape and sorting, as well as diversity, depend on the nature of the original associations as well as hydrodynamic processes. What are important are sequential changes (vertically and laterally) in these parameters. Figure 7.7 shows flash cards to give an assessment of a skeletal accumulation, comprising shells, crinoid elements and ooids (or rounded grains), and which can be used separately or in combination. The matrix should be described separately.

7.2.6   Other aspects

For the sedimentologist there are also four other *aspects* that, though bearing more on the palaeontological information, are nevertheless pertinent:

1. the taphonomic history of the material (Chapter 4),
2. the composition and diversity (monotypic, polytypic) present (§6.2.3.7, 6.2.3.8),
3. lateral and vertical changes in composition,
4. composition in different lithologies.

The sedimentologist is concerned with assessing distance and process of transport, and sediment source. Assess relative breakage (§6.2.1) and roundness, with respect to composition, skeletal encrustation (§6.2.1) and degradation, possibly indicating prolonged exposure on the sea floor (§7.1), variation in preservation mode (§8.4.2f) that might indicate intraformational reworking, or derivation from much older sediments. Fossiliferous pebbles may also provide clues to source area.

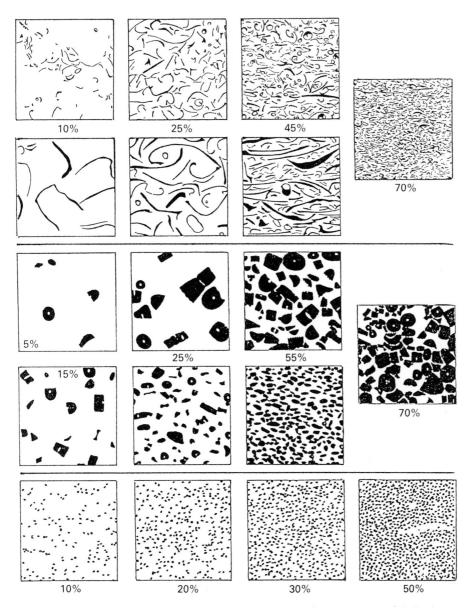

Figure 7.7 Semiquantitative comparison diagrams for estimating percentage of skeletal material (shells, crinoid elements) and ooids in a sediment. *To be used at ×2* (redrawn and abbreviated from Schäfer, 1969)

7.3   Bioturbation and trace fossils (Table 7.5)

Bioturbation is the reworking of sediment by an organism or organisms. It means disturbance of the primary sedimentary structures and grain fabric (formed by current, wave or wind action), thus imparting a new fabric. There is usually change in the grain size distribution, and therefore in porosity and

Table 7.5 The effects and implications of bioturbation

Observations	Implications
Lamination destroyed, grain orientation destroyed, coarse and fine laminae mixed	an averaging of overall grain size; overall reduction of porosity, and reduced permeability; sediment maintains higher water content lithification delayed
Fines selectively removed and passed into suspension	increase in grain size, porosity and permeability
Fines moulded into faecal pellets	sediment grain size effectively increased
Sediment trapped by tubes or plant stems	decrease in grain size; stabilization of substrate
Sediment bound by roots	stabilization of sediment
Vertical burrows	act as drainage channels (hence, sediment firmer than expected); permeability normal to bedding increased
Bedding parallel burrows	permeability normal to bedding probably reduced; permeability parallel to bedding may be enhanced
Burrow-wall lined	preferred permeability pathways
Open burrows vastly extending sediment – water interface	substantially increased geochemical fluxes below substrate

permeability. Bioturbation can be relatively minor, leading to a mottled sediment or a few distinct trace fossils, or can be so extensive that nothing remains of the primary stratification. To state that a sediment is bioturbated is like saying that a sediment is fossiliferous: it does not convey much. The degree of bioturbation can be quantified or graded (Fig. 7.8; *see also* §6.2.4.1). Other schemes that are in common use are: slight, moderate, intense bioturbation (often used in logging), and Howard & Reineck's (1972) scheme determined by X-radiography of box-cores taken from siliciclastic shelf sediments:

grade	0	no bioturbation
	1	trace of bioturbation
	2	<30%
	3	30–60%
	4	60–90%
	5	90–99%
	6	100%

As with any sedimentary structure, some variability within the sediment, in mineralogy, colour or grain size, is necessary for bioturbation to be visually recognizable. Mud between sand beds that has been piped into the sand, or *vice versa*, is ideal, but well-sorted quartz sand can present problems. What is to be done if beds of sand display neither primary sedimentary structures nor indication of bioturbation? Do not assume that they are bioturbated. Sediments can rarely become homogenized by biogenic processes, since the final burrows will always be present. X-radiography often reveals structures virtually invisible to

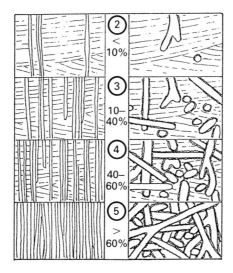

Figure 7.8 Flash-diagram to assess degree of bioturbation for *Skolithos* (left) and *Ophiomorpha* (right). Each frame represents an area 50 × 35 cm normal to bedding. For other types of ichnofabric separate diagrams must be constructed (after Droser & Bottjer 1990). (Grade 1 – no bioturbation): Note that no indication of tiering is expressed, *see also* Fig. 7.24

the eye. Be prepared to collect oriented specimens for slabbing. There are three aspects of ichnology that are most useful for the sedimentologist:

1. The *relationships* between the distinct biogenic structures (trace fossils) and the bedding, and to the fabric and texture of the sediment. Appreciation of these aids in understanding of sedimentological processes and their timing. (Discussion of these relationships forms the main part of this section.)
2. The *environmental significance* of the trace fossils. Although this is the classic and best-known sedimentological application, the basic factors involved are ecological (§6.2.4).
3. The *ichnofabric*, which combines the above two (§7.3.6).

There are two approaches to bioturbated sediments: to investigate the bioturbation and trace fossils that have replaced the primary structures; or, to attempt to determine what the primary structures originally were i.e attempt to 'restratify' the sediment. The question to be asked is what did the sediment look like before it was bioturbated?

7.3.1 Major 'event' sedimentation

In major 'event' sedimentation (turbidites, storm events) trace fossils provide convincing confirmation of an event which interrupted infaunal life. In siliciclastic turbidites and storm units, the sequence of sedimentary structures (Fig.

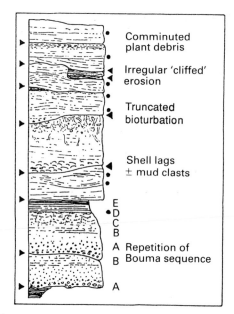

Figure 7.9 Criteria for recognizing amalgamation in turbidites (below) and storm beds (above). Arrows indicate 1st order discontinuities at amalgamations; spots, 2nd order discontinuities within or between flow regimes

7.9) is generally indicative, but this can be unclear, especially in carbonate deposits. If it is clear that colonization took place only from the top of the bed and not from any level within, then there can have been only a single phase of sedimentation. Further confirmation in turbidites, if required, can be by analysis of the suite of sole traces associated with turbidites of different thickness. In thick event beds, only traces made by animals able to penetrate right to the sole, will be found. Take care to exclude amalgamated units (Fig. 7.9).

Turbidites Trace fossils, although often distinct and diverse on the soles of turbidites, are usually few in number. Assess the assemblage present over a sequence. The best way is to record on a log, noting whether the traces occur

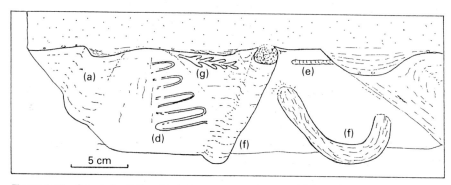

Figure 7.10 Criteria to distinguish between pre- and post-turbidite suites of trace fossils on turbidite soles (for explanation see text). Scale bar = 5 cm

on the sole or top of each bed. The sole traces represent two suites (Fig. 7.10; Kern, 1980): those etched out by the turbidity current from the background mud and then infilled (pre-turbidite suite) (Figs 7.11, 7.12); and those which colonized from the top of the bed (post-turbidite suite). Note that in most turbidites it is rare to see burrows traversing a bed because of the nature of the sediment.

Criteria for distinguishing the pre-event suite are:

(a) Erosional modification of traces by the turbidity current (Fig. 7.10a).
(b) Displacement of faecal strings.
(c) Traces on soles of very thick beds are likely to be pre-depositional, but beware of amalgamated beds.
(d) Partial erosion of laterally extensive traces (Fig. 7.10d).
(e) One would expect that cross-cutting relationships would be indicative: e.g. where a flute cast cuts a burrow. But, if the groove is deep and sharply bounded, then the producer could have cut straight through the groove, though this is unlikely (Fig. 7.10e).

For post-depositional traces:

(f) Modification of sand-mud interface by trace or detachment of the burrow from the sole (Fig. 7.10f).
(g) Trace conforms to shallow groove cast (Fig. 7.10g).
(h) Grooves deflected slightly by burrow.
(i) Cross-cutting relationship (above: e) must be used with care.
(j) Large traces unaffected by current action.
   NB: Care must be exercised where load casts have formed.

The tops of turbidites are often intensely bioturbated, especially in the late Mesozoic and Caenozoic, following the rapid increase in diversity of the post-event suite at the end of the Cretaceous. Search for distinct traces. If the study area is sufficiently large, then the turbidites may show lateral changes in sedimentary character, proximal to distal. There may also be changes in the vertical sequence, e.g. associated with a coarsening upward sequence.

In *storm beds* (Figs 7.13–14), bioturbation is often less distinctive when compared with turbidites, and there is generally less difference between the pre-event and post-event suite. Bioturbation may be pervasive, rendering the sharp sole indistinct, due to extensive and deep *Ophiomorpha* and *Thalassinoides* burrows. Interbeds between events are also typically bioturbated. Again, be on the lookout for amalgamation. Incomplete trace fossils can give an indication of the amount of degradation (Fig. 7.15).

Tidal and tidal-influenced facies (Fig. 7.16) High energy sequences of cross-stratified sands and oolites seldom display trace fossils. If bioturbation can be found it is likely to be in two settings: (1) along the 'bottom set' indicating colonization between dunes and, (2) extending upwards along the

Figure 7.11 *Paleodictyon minimum* as pre-depositional convex (positive) hypichnion on sole of turbidite (Eastern Turkey, Eocene). Scale bar = 1 cm

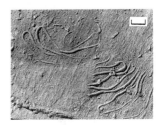

Figure 7.12 Turbidite sole with partly eroded pre-turbidite trace (positive hypichnion) *Helminthoida crassa*, and post-depositional trace *Strobiloraphe clavata* (Zollhaus nr. Fribourg, Switzerland, Palaeocene). Scale bar = 1 cm

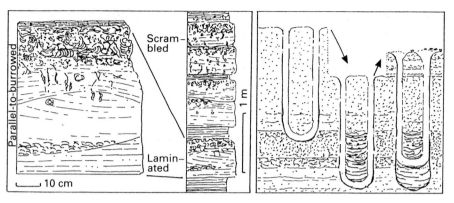

Figure 7.13   (Left) Typical bioturbation in shelf storm beds (either siliciclastic or calcareous): parallel (laminated)-to-burrowed (lam – scram). (Right) U-burrows with adjustment to small amounts of aggradation or degradation forming protrusive and retrusive spreiten (Fig. 7.18)

Figure 7.14   Neritic sandstones displaying relatively sharp soles and bioturbated tops: parallel-to-burrowed (Upper Jurassic, northern France).

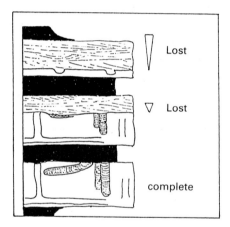

Figure 7.15   Trace fossils as indicators of amount of sediment lost by penecontemporaneous erosion (after Aigner, 1985)

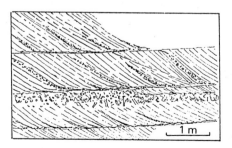

Figure 7.16   Bioturbation intervals in bottom set, and at intervals of slack water in tidal cross-stratification

cross-stratifications indicating pauses in advance of the dune front (important in assessing nature of tidal cycles etc.).

Owing to the rapid changes that take place in high energy tidal channels, keep a lookout for chance preservation of a marine trace fossil. This may be important in deciding whether one is dealing with a fluvial or estuarine environment. But note that only an ephemeral event, such as when a salt wedge has extended into a distributary system, may be represented.

Ashfall facies   Deposition of an ash is often disastrous for the biota in aquatic environments: benthos, plankton and nekton. Anticipate that there may be much biologically relevant information to be gleaned below the ash.

7.3.2   Minor event sedimentation (Fig. 7.13)

In shelf and shallow water sediments, minor events of sedimentation and erosion commonly occur owing to fluctuating energy conditions. Organisms can often adjust to such events and, because so many trace fossils were formed at a high angle to the substrate surface in shallow water, it is easy to detect vertical adjustment or truncation. Adjustment of a few mm–cm at any one stage can be accomplished by many bivalves, mobile echinoderms, anemones as well as annelids and arthropods. Only annelids, arthropods and gastropods can escape upwards through an appreciable sediment cover.

Tidal flats adjacent to a vigorous channel may show little bioturbation because of the high frequency of small-scale aggradation and degradation. In quieter areas, the likelihood of the sediment becoming bioturbated is much greater.

7.3.3   Nature of substrate (Fig. 7.17)

Was the ancient substrate surface soft, loose, firm or a lithified hardground or rockground? Does it represent a discontinuity surface? If so, could it be used in correlation (Table 8.1)? Autochthonous fossils generally give a good indication of the nature of the substrate at the time of colonization. Gradual hardening can be reflected in a succession of colonizers responding to the changes (§6.2.4). If emersion, with lithification, interrupted the succession, soft substrate traces will be overprinted directly by borers and encrusters. The situation can be complex, and it may be necessary to collect for later analysis to determine the cementation history. As a general rule coarser grained sediments (oolites, skeletal sands) cemented quickly: too quickly for a 'gradual' biogenic succession to form.

A. *Criteria for a loose or soft substrate* (adapted from Gruszczynski, 1986; Goldring and Kazmierczak, 1974):
1. Indefinite burrows with irregular outline (mm–cm diameter);
2. Bivalve burrows associated with adjustment or escape features (Fig. 6.20);

Figure 7.18 *Diplocraterion* 'yoyo', (left) protrusive form with complete U-burrow, (centre) upper part of protrusive form, (right) retrusive form. All three were truncated prior to deposition of capping sand (Upper Devonian, North Devon, England). Scale bar = 1 cm

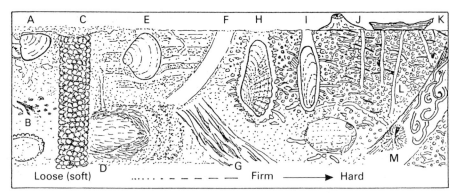

Figure 7.17 Diagram to show types of biogenic response to loose (soft), firm and hard (lithified) substrates: (approx. ×0.5 unless otherwise indicated): (A) *Nucula*; (B) *Chondrites*; (C) *Ophiomorpha*; (D) *Echinocardium*; (E) *Mercenaria/Arctica* (×0.25); (F) *Thalassinoides*; (G) *Spongeliomorpha* (with scratched burrow wall); (H) *Petricola*; (I) *Lithophaga* (insoluble grain on left side) (×1.0); (J) crinoid holdfast; (K) oyster; (L) *Trypanites*; (M) encrusting bryozoan and serpulids on cemented burrow wall

3. U-shaped burrows with lined wall, or annular zone;
4. *Ophiomorpha* burrows with pellet-stabilized walls (Fig. 6.22);
5. Burrows made by spatangid echinoids e.g. many *Scolicia*;
6. U-shaped burrows with spreite (*Diplocraterion*) (Fig. 7.18);
7. Originally upright now compacted burrows.

B. *Firm substrates*:
1. *Spongeliomorpha* (*Thalassinoides paradoxicus*) (Fig. 7.17) with scratch marks on walls;
2. Bow-form *Arenicolites* (without spreite);
3. Various boring bivalves e.g. *Petricola*, *Zirfaea*;
4. Encrusters e.g. *Liostrea* on firm to hard substrates in the Jurassic.

C. *Hard substrates* (commonly with mineralized crust):
1. Bivalves boring by mechanical means, *Pholas*, *Barnea*, producing the boring *Gastrochaenolites* (Figs 6.23, 7.21);
2. Bivalves boring by (mainly) chemical means *Lithophaga*, *Gastrochaena*, also producing *Gastrochaenolites*;
3. Borings made by regular echinoids;
4. Small borings made by annelids, sponges, bryozoans producing such trace fossils as *Trypanites* (Fig. 6.23), *Entobia* (Fig. 6.6);
5. Encrusting oysters e.g. *Nanogyra* in Jurassic;
6. Encrusting bryozoans, serpulids, sponges and corals (Figs 7.20, 7.21);
7. Crinoid holdfasts (Fig. 7.19).

Figure 7.19 Crinoid holdfasts and oysters attached to hardground (Middle Jurassic, Bradford-on-Avon, England). Scale bar = 1 cm

Borings in grainstone can generally be proven with the aid of a hand lens: because ooids, shell fragments etc. are truncated evenly with the cemented matrix. This is less easy to demonstrate in fine grained sediments. Many ancient muddy sediments must have been very soft and 'soupy' in the top few mm or cm. Animals must swim rather than burrow through such sediment, but may form stabilized or constructed walls. High compaction and smearing of burrows indicate an original high water content. Below, down to about 30 cm in modern pelagic sediments, the sediment is firm, and it is in this zone that most of the burrows that are eventually preserved, are formed. Absence of the uppermost, mixed-layer commonly reflects its loss by penecontemporaneous erosion.

In muddy sediments, pose the question: was the sediment originally pelleted and thus behaving more like sand than mud? Search for early formed concretions with 'frozen' pellets or in shelter cavities (Fig. 4.19) where the pellets have been unaffected by compaction. Some early hardening of micritic pellets was probably necessary for their preservation. Micro-organisms commonly form a film on the sediment surface, the former presence of which may be deduced from riffling of tool marks and creasing around fossils. Such films hindered penecontemporaneous erosion.

7.3.4 Chemical events: changes in oxygenation or salinity

The presence of fine laminations in dark, argillaceous and pyritic sediment indicates anaerobic conditions. Sequences of such sediment, punctuated by thin bioturbated horizons (generally *Chondrites*) (Figs 6.12, 6.13), probably indicate temporary intervals where the redox horizon dropped below the substrate surface. In estuarine or delta-mouth facies, seasonal shifts in the saline wedge may lead to temporary colonizations.

7.3.5 Grain-size distribution, porosity and permeability

A grain-size distribution derived from bioturbated sediment can have no hydraulic significance. Likewise, bioturbation, in destroying primary current structures and the fabric, also destroys the directional permeability associated with the structures. Strongly bioturbated intervals will normally have reduced permeability, but the mixing-in of a proportion of mud seems often to have hindered cementation. A relatively high porosity is thus retained. Sand-filled burrows penetrating heterolithic sediments can act as conduits between the muddy layers. In general, vertical burrows may increase vertical flow, and horizontal burrows, horizontal flow. Burrows formed at depth in consolidated sediment may have a loose fill, or even remain open (Fig. 7.22). This much enhances permeability. In partly bioturbated sediments primary grain-size distribution can be measured, if care is taken to sample only areas exhibiting primary sedimentary structures. (In lithified rock this can be done on a thin-section: 50–100 grains may be sufficient.)

Figure 7.20 Hardground encrusted by *Aulopora*, and bored by Trypanites (Devonian, Iowa). Scale bar = 1 cm

Figure 7.21 Oyster-encrusted hardground with opening to borings of *Trypanites* and *Gastrochaenolites* (Middle Jurassic, near Cirencester, England). Scale bar = 1 cm

Figure 7.22 Rock below a hardground where burrows have remained open (Middle Jurassic, Gloucestershire, England). Scale bar = 1 cm

7.3.6   Ichnofabric, and trace fossils in cores (Fig. 6.23)

So often, at outcrop or in cores, the sedimentologist is faced with bioturbated sediment with a minimum of primary sedimentary structures preserved. What can be done? The primary fabric is well-nigh lost and a new fabric imparted. This is the result of animal activity on all scales (and of course roots in soils). The situation can be so serious that one is tempted to turn away in dismay, but it is important to try and understand how it all came about. And, since ichnofabric integrates the diagenetic, stratinomic and ethological (Fig. 6.20) aspects with the sedimentology, it is likely to be significant. Trace fossils in cores present special difficulties, in that it may not be easy to see the nature of the bioturbation on the rough outside or sawn surfaces, or because of the film of drilling mud. (Note that this mud may penetrate the core.) Cores may be oblique to bedding. Actual bedding information may be difficult to discern. But there is often the advantage that muddy and sandy sediment core equally well. In cores, or even at the exposure, it is often not possible to identify traces to generic level. They can be categorized on a few parameters, which can be tabulated and programmed.

Information that might prove useful includes (*see also* §6.2.4):

1. Tiering (Fig. 7.23 and Table 7.6) of burrows is clearest below the upper surface of event beds (§7.3.1) in the post-event suite. It is seldom simple to distinguish shallow-tier from deep-tier traces. (Refer to Figs 6.19, 6.23, 6.24.) Traces that are normally in the shallow tier (e.g. *Cruziana*, *Rhizocorallium*, *Planolites*, *Aulichnites*) can extend more deeply in reworking the fill of other burrows (e.g. *Planolites* in *Thalassinoides*).
2. Ordering (the sequence of burrows) is clearest at omission surfaces, but can also be seen at an environmental (facies) change, or indeed, where a change in the trace-making population occurred possibly by chance. An ordering is also evident where sediment accretion has brought deeper-formed burrows against shallower-formed burrows. In this situation it may be difficult to unravel the original tiering that was

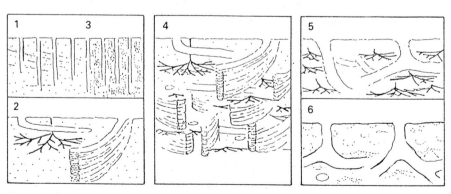

Figure 7.23 Six styles of trace fossil tiering (for further explanation see text)

Table 7.6 Tiering and ichnofabric

	Observations	Implications	Environmental stability	Porosity/permeability (*p/p*)
1.	Single colonization with primary structures clear	colonization of an event bed, (aggradational event) by low diversity biota	generally associated with high stress environment e.g. estuarine	enhancement of permeability normal to bedding common
2.	Simple tiering	colonization of an event bed by tiered suite	typical of wider range of environments; associated with event beds	as above, or reduced *p/p* at bioturbated level
3.	Single successive colonization	essentially gradual aggradation in low diversity (stressed) environment	relatively higher than 1,2	depends on type of trace
4.	Complex tiering	gradual aggradation of tiered suite	stable environment	porosity/permeability reduced
5.	Gradational overlap	gradual environmental change		as above
6.	Omission overlap	sharp facies change ± hardground; often associated with some condensation		generally associated with *p/p* break occasionally 'open' maze/boxwork enhances *p/p*

present (Fig. 7.24). Ordering is generally best analysed on bedding surfaces.

3. Attempt to ascertain the degree of bioturbation and frequency of each ichnotaxon for specific facies (above).
4. Size, outline and attitude of burrows, type of branching. A general assessment of the relative abundance of vertical and horizontal burrows is a useful step, since vertical burrows are most typical of sandy shallow-water environments (where they are typically made by suspension feeders), in contrast to burrows made by deposit feeders in muddier sediment, in more stable and often deeper-water environments.
5. Presence and type of wall lining (smooth or pelletal film, shell fragments, mud, zoned).
6. Type of burrow fill (meniscate, symmetrical/asymmetrical, annular, pelleted, unstructured), and manner of filling (more difficult) (passively, by sedimentation, actively by organism e.g. faecally, by collapse or by compaction).
7. Diversity of burrow types.
8. Vertical change in degree of bioturbation, burrow types and in diversity.

Slabbing may provide more information. Consider carefully before cutting, and try to anticipate what may be revealed. For traces normal to bedding, slabbing should be parallel to bedding. Thus, to distinguish a meniscate burrow

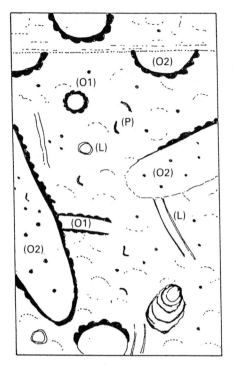

Figure 7.24 A simple ichnofabric in which pellet-lined burrows (*Ophiomorpha*) and mud-filled, unlined burrows (*Planolites*) are characteristic. The ordering is somewhat uncertain: (1) background mottling; (2) small *Ophiomorpha* (O1), lined burrows (L); (3) large *Ophiomorpha* (O2); (4) mud-filled burrows *Planolites* (P) which penetrate *Ophiomorpha* as well as being common through the fabric. Typical of middle shoreface sands of the Mesozoic–Caenozoic e.g. Upper Jurassic North Sea; Eocene, southern England. The sharp truncation of *Ophiomorpha* at the top suggests substantial penecontemporaneous erosion. Note that *Ophiomorpha* may be roof-lined only, or pass into an unlined burrow. Bioturbation grade (Fig. 7.8): 3/5 (distinct burrows (grade 3) superimposed on completely bioturbated sediment (grade 5) × 0.5

as a retrusive *Diplocraterion* and not *Teichichnus*, the cut should be made perpendicular to the direction of retrusion (Fig. 6.23). In ichnofabric analysis it is important to appreciate (and describe and figure) sections both normal and parallel to the stratification (or lamination).

7.4   Tool marks

Tool marks (Fig. 7.25) are the result of objects such as mud clasts, together with skeletal and plant material (fragmentary or in more complete condition), being rolled or saltated by hydraulic processes (current or wave action) over the substrate surface and locally impinging on the surface. The marks frequently leave a sense of flow direction or wave trend and thus may be useful when other indications are absent. Tool marks are seen on upper surfaces and as casts on soles. They may be grouped as roll or tumble marks, brush marks (elongated

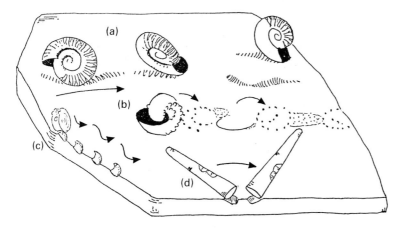

Figure 7.25 Tool marks. (a) roll marks made by a swaying perisphinctid ammonite on a fine-grained substrate; (b) tumbling marks made by an inflated, tuberculate ammonite; (c) skip marks made by a vertebra; (d) orthocone prod mark. (a, b after Seilacher 1963, c, d after Dzutynski and Walton 1965)

with small mound at down-current end), bounce marks (symmetrical), prod marks (asymmetrical, broadening down-current), skip marks (repeated impact) and drag marks (much elongated). Wave action on a rooted weed can lead to circular marks (Fig. 5.1). Only occasionally is it possible to identify the exact nature of the tool. The high degree of preferred orientation means that tool marks are seldom confused with trace fossils (above and §6.2.4). Biological objects such as the tubes of emergent worms may interfere with and destabilize current flow, leading to local scouring. Pressing a piece of soft Plasticine onto casts helps in appreciation of the mark.

References

AIGNER T 1985 *Storm Depositional Systems* Springer, 174 pp

DROSER M L, BOTTJER D J 1987 Development of inchnofabric indices for strata deposited in high-energy nearshore terrigenous clastic environments. *In* Bottjer D J (ed) *New Concepts in the Use of Biogenic Sedimentary Structures for Paleoenvironmental Interpretation.* Society of Economic Paleontologists and Mineralogists, Pacific Section, p. 29–33

DROSER M L BOTTJER D J 1990 Ichnofabric of sandstones deposited in high-energy nearshore environments: measurement and utilization *Palaios* 4: 598–604

DZULYNSKI S, WALTON E K 1965 *Sedimentary Features of Flysch and Greywackes* Elsevier, 274 pp

FÜRSICH F T, OSCHMANN W 1986 Storm shell beds of *Nanogyra virgula* in the upper Jurassic of France. *Neues Jahrbuch für Geologie und Paläontologie Abhandlungen* **172**: 141–61

FUTTERER E 1982 Experiments on the distribution of wave and current influenced shell accumulations. *In* Einsele G, Seilacher A (eds) *Cyclic and Event Stratification* Springer, p. 175–9

GOLDRING R, KAZMIERCZAK J 1974 Ecological succession in intraformational hardground formation. *Palaeontology* **17**: 949–62

GRUSZCZYNSKI M 1986 Hardgrounds and ecological succession in the light of early diagenesis (Jurassic, Holy Cross Mountains, Poland) *Acta Palaeontologica Polonica* **31**: 163–212

HOWARD J D, REINECK H-E 1972 Georgia coastal region, Sepelo Island, USA: sedimentology and biology. *Senckenbergiana maritima* **4**: 81–123

KERN J P 1980 Origin of trace fossils in Polish Carpathian flysch. *Lethaia* **13**: 347–62

KIDWELL S M, FÜRSICH F T, AIGNER T 1986 Conceptual framework for the analysis and classification of fossil concentrations. *Palaios* **1**: 228–38

SCHÄFER K 1969 Vergleichs-Schaubilder zur Bestimmung des Allochemgehalt bioklastischer Karbonatgesteine. *Neues Jarhbuch für Geologie und Paläontologie, Monatshefte* p. 173–84

SCOTT A C 1979 The ecology of Coal Measure floras from northern Britain. *Proceedings of the Geologists' Association* **90**: 97–116

SEILACHER A 1963 Umlagerung und Rolltransport von Cephalopoden-Gehäusen. *Neues Jahrbuch für Geologie und Paläontologie, Monatshefte* p. 593–615

Texts on sedimentary environments and facies

BATHURST R G C 1975 *Carbonate Sediments and Their Diagenesis* Elsevier, 658 pp

COLLINSON J D, THOMPSON D B 1989 *Sedimentary Structures* (2nd edn), George Allen & Unwin, 207 pp

EINSELE G, SEILACHER A (eds) 1982 *Cyclic and Event Stratification* Springer, 536 pp (Numerous papers and well illustrated)

FÜCHTBAUER H (ed) 1988 *Sedimente und Sedimentgesteine* (4th edn) Schweizerbart' sche, Stuttgart, 1141 pp

LEEDER M R 1982 *Sedimentology, Process and Product* George Allen & Unwin, 344 pp

PETTIJOHN F J 1975 *Sedimentary Rocks* Harper & Row, 628 pp

READING H G 1986 *Sedimentary Environments and Facies* Blackwell, 615 pp

REINECK H-E, SINGH I B 1980 *Depositional Sedimentary Environments* Springer, 549 pp

SCHOLLE P A BENOUT D G, MOORE C H (eds) 1983 *Carbonate Depositional Environments*. American Association of Petroleum Geologists Memoir No. **33**, 708 pp

SCOFFIN T P 1987 *An Introduction to Carbonate Sediments and Rocks* Blackie, 274 pp

TOOMEY D F (ed) 1981 *European Fossil Reef Models* Society of Economic Paleontologists and Mineralogists, Special Publication **30**, 546 pp

TUCKER M E 1981 *Sedimentary Petrology: An Introduction* Blackwell, 252 pp

TUCKER M E, WRIGHT V P 1990 *Carbonate Sedimentology* Blackwell, 496 pp

WALKER R G (ed) 1979 *Facies models* Geoscience Reprint Series 211 pp

WILSON J L 1975 *Carbonate Facies in Geological History* Springer 471 pp

8

Fossils for the stratigrapher

'There is no doubt that a major restriction on (sedimentological) research is the inadequacy of stratigraphic correlation', E. Mutti speaking at the de la Beche Symposium, Imperial College, London, February 1986.

8.1 Introduction

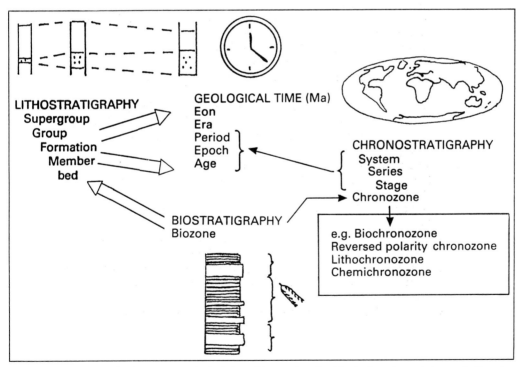

Figure 8.1 Relationships between geological time, lithostratigraphy, biostratigraphy and chronostratigraphy

Until outcrops, sections and wells are correlated by time lines, there is no way of gaining any real appreciation of the temporal distribution of past environments across an area or within adjacent basins and ranges; let alone of clarifying what was going on at distant points on the globe. Further, until the rocks are dated, it is not possible to get far in understanding the structural development of the area.

Through much of the Phanerozoic, and in a general way also in the Proterozoic, fossils offer the most precise means of correlation available. There are several methods of correlation other than by using fossils, and Table 8.1 shows some of these with the service area for each, and the readiness with which each may be used in the field. Notice that many of the lithological methods are more distinctive than palaeontological methods, and that there are many methods that can be used for a small area. Also, fossils do not provide the only means of correlating on an international (chronostratigraphic) scale (Fig. 8.2). No mention is made in Table 8.1 of geochronology (absolute dating) using radiometric methods, because it is not applicable for a field investigation and it is the least precise, though it does, of course, provide a framework for litho-, bio-, and chronostratigraphy. Be on the lookout, therefore, for rocks, such as tuffs, and those containing autochthonous glauconite that might be radiometrically dated.

Every fossil has some stratigraphic value, so that the smallest sponge spicule or tiny lingulid brachiopod is indicative of the Phanerozoic rather than the Precambrian (and this may be important). It is useful to define the degree of stratigraphic refinement required for the particular job in hand. There are approximately four orders of magnitude of stratigraphic refinement (Table 8.2).

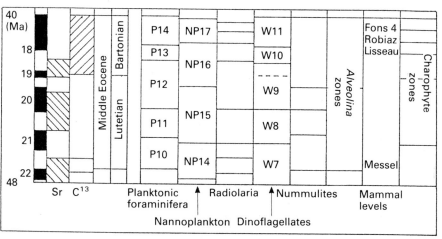

Figure 8.2 An example of a correlation chart: the Middle Eocene, Lutetian and Bartonian, showing zonal schemes for various groups of fossils against radiometric scale (Ma), magnetic anomalies 19–22, and chemizones marking changes in isotopic composition, Strontium (Sr) and Carbon (C¹³) (after Cavelier & Pomerol, 1986)

Table 8.1 Some features of correlative value, with distinctiveness in field and usual geographical extent

Feature	Field identification	Normal range (km^2)	Comments
Lithological			
turbidite event bed	poor to good	<10^4	amalgamation may reduce correlative value may be discontinuous
sheet sandstone	poor to moderate	<10^2	
soil horizon (palaeosol)	good	<10^3	
thick coal seam	moderate	<10^3	splitting may make identification difficult
marine black mudstone	good	<10^4	often excellent for widespread correlation when linked with biota
local cycle of lithology	distinctive	<10^4	care required in correlation when many cycles present
shell horizons, bone beds	distinctive	<10^2	use with care, often unreliable
hardgrounds	generally good	<10^4	may be local, useful if grouped into a pattern
sequence boundary	generally good	10^4–10^6	local to international
Volcanic event			
tuff horizon	good to very good	<10^4	especially useful if grouped
Extraterrestrial			
irridium event	—	>10^6	should be world-wide, but needs specialist equipment to identify
Milankovich cycles	poor to good(?)	>10^6	not necessarily world wide, and needs careful analysis
Faunal/floral			
plant, coral, brachiopod taxa	poor to good	<10^4	relatively easy to use with a suitable reference text for local biostratigraphic correlation
ammonoid, graptolite trilobite zonal taxa	poor to good	>10^6	often of international application
spore, foraminifera, coccolith zonal taxa	too small for field ident.	>10^6	
Geophysical			
seismic reflector	—	<10^4	generally of basinal application
down-hole log	—	<10^4	generally of basinal application
magnetic reversal	—	>10^6	chronostratigraphic application
magnetic susceptibility gamma-ray spectrography	can be good	<10^2	best in siliciclastic sediments
Geochemical			
isotopic peaks (e.g. 0^{18} max.)	—	>10^6	best in siliciclastic sediments

Table 8.2 Stratigraphic refinement

1000 my	(650 my)	Phanerozoic
100 my		System level: Devonian
10 my	(5–10 my)	Stage or series level: Palaeocene Series, Kimmeridgian Stage
1 my	(0.5–5 my)	Zone level: *Didymograptus bifidus* Biozone

At system and series level, one is generally concerned with the range of a genus or higher taxonomic category. There are a number of older but well illustrated texts that are useful at this level for identification, if it is appreciated that advances will have been made since their publication in definition, nomenclature and, possibly, actual ranges cited.

Determination to zone level requires an appreciable amount of knowledge about the biota, and generally this is a job for the specialist. However, it is as well to know what is required. A really useful stratigraphic fossil needs five qualities: (a) it should be fairly common, (b) it must be widespread, on an international scale, (c) it should be present in many facies, (d) its stratigraphic range should be narrow, and (e) there must be good description (with illustration) of the species. There are few organisms that can fulfil all these qualifications, and many monographs do not adequately describe the species or cover the palaeoecology. Nevertheless, try to find out about the biostratigraphy of the area it is intended to visit in order to get an appreciation of just how good the biostratigraphy is, and to get the most from the fossils that will be collected.

Collecting fossils can take time. It is important that the lithological succession is established in some detail, so that the stratigraphic level of the fossils is known accurately. The primary objectives in mapping and well-logging are to establish the stratigraphical succession and distribution of lithologies. This is followed by correlation of the stratigraphical units within and beyond the area under investigation. For field techniques and procedures in mapping, see references in Chapter 3.

8.2   Primary lithostratigraphy and facies analysis

Describing sedimentary rocks needs clarity and precision. For each member and formation that is recognized:

1. designate an accessible type section;
2. identify and describe the lower boundary precisely (the upper boundary is taken care of by the succeeding unit);
3. provide an illustrated description of the lithologies, including any lateral changes;
4. identify and describe diachronous facies;
5. identify and describe diagenetic facies (e.g. secondary dolomite);
6. determine the thickness and lateral extent;
7. determine the age.

Putting lines onto a map is always an exciting moment. Formation boundaries represent changes in lithology (Fig. 8.3). Some, like unconformities and disconformities, are generally distinct, but both can be indistinct where the underlying sediment was reworked biologically, or where scree follows a soil profile, as at a landscape unconformity. Within a conformable succession,

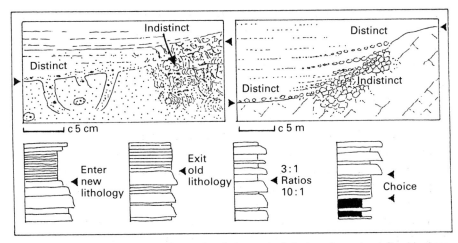

Figure 8.3 Lithological boundaries can be distinct or indistinct, and may be defined in three principle ways

Table 8.3 Summary of general strategy

Preparation (Chapter 3)
Reconnaissance of area, cores
Lithological mapping
Establish lithostratigraphy
Collect and sample for fossils and microfossils (Chapter 3)
Establish biostratigraphy
Integrate with regional stratigraphy

---

there are just three types of boundary:

1. the *appearance* of a new lithology such as the introduction of a black, laminated shale into a sequence of muddy shales and sandstones;
2. the *termination* of a lithology such as the termination of graded greywackes in a sequence in which they are prominent, but with continuation of shales;
3. the *gradual change* in lithological proportions, such as sandstone–shale ratio. Here there appears to be no significant event which led to a facies change, but when the succession is viewed overall, there is an apparent need to divide the succession, if only to make for more manageable units. An arbitrary decision will have to be made e.g. on the sandstone : shale ratio.

There can be a conflict of interest on boundaries between the requirements of the map maker and the interests of the sedimentologist or palaeontologist. In Fig. 8.3 (right) there is a choice of boundary, which may be placed either at the appearance of thick sandstones or at the disappearance of black shales. A boundary placed at the appearance of the thick sandstones is likely to be more easily mapped, but some might argue that the disappearance of black shales might be taken as a more significant event.

Facies analysis is the most difficult part of a sedimentological analysis. In summary, one has to try to distinguish the various lithologies and group them. The first stage is to log sequences (§3.4). The bed is the basic unit (§3.3, Fig. 3.1). It is characterized by its lithology, sedimentary structures, common fossils, and vertical and lateral changes in these. Sometimes a bed can be unique to a sequence, thus constituting a facies by right. Generally, beds can be grouped into facies. Many, if not most, sedimentary successions comprise either sequences of facies, rhythms and cycles (§3.3), or event beds including turbidites, storm beds, ash falls (§7.3.1) or these in combination. Sequences of facies reflect changing environments, and establishing sequences is the first stage of basin analysis. Each formation boundary will coincide with a particularly significant facies boundary. The old dictum that 'boundaries are drawn in the field and not in the pub' applies to formation boundaries and to facies boundaries. The only sure way is to walk them out. Note that the distribution of fossils may be mapped independently of the distribution of lithologies, though this is done but rarely on a scale greater than about 1 : 10,000. Fossils closely associated with particular facies are referred to as facies fossils (§8.4.3).

Conflict can arise over the systematics of facies analysis. Recognizing and distinguishing fossil species and sedimentary facies are similar operations. The reputations of palaeontologists and sedimentologists depend on how well they are carried out. If we agree that a sedimentary facies represents the total makeup of a sediment, then problems can arise if the biological components change at a different rate to the hydraulically or chemically determined features. For instance (Fig. 8.4), in a lagoonal sequence, repetitive cycles of four

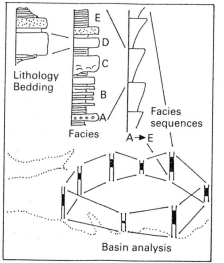
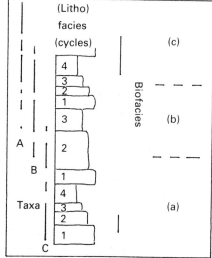

Figure 8.4   (Left) Basin analysis begins with the analysis of a bed. Then follows facies recognition and analysis of sequences of facies, before basin analysis can be undertaken. (Right) Relationships between lithofacies and biofacies (see text)

distinctive lithologies contain a rather sparse biota, but one that changes from dominantly freshwater at the base to near-marine at the top. It is generally better to recognize the four (litho)facies and to describe the change in biota independently.

Facies changes can be abrupt, gradational or by intercalation. Gradational changes are frequently due to bioturbation and were originally abrupt. Change in the proportion of an intercalating lithology e.g. shales in sandstones can be quantified.

8.3　Biostratigraphy

How does one know which fossils will be useful in correlation? There is no simple answer to this question, and much will depend on what preparation has been done, and on the age of the strata (Table 8.4).
Biostratigraphy is the organization of sedimentary sequences based on biological events (Figs 8.5, 8.6): appearances, disappearances and, less commonly, acmes of taxa (change in a ratio can also be used). Changes in the fossil assemblages through a more or less continuous sequence can be attributed to four possible natural causes:

1. changes in ecological tolerance (for example substrate preference, salinity, temperature, turbulence);
2. migration, for instance, where changes in palaeogeography have allowed land animals to migrate, or marine organisms to increase their geographical distribution;
3. evolution;
4. extinction.

In 1 and 2, further ecological change may have resulted in the taxon (taxa) reappearing higher in the stratigraphy: the Lazarus effect. 'Evolution' can also appear to be an extinction (pseudoextinction) when faunal or floral lists are examined, because of name change(s) at genus or higher level coinciding with a chronostratigraphic boundary (generally system boundary). Extinction at genus or higher level must affect the whole clade. There are two other aspects that are, at least as, if not more, important: (1) preservational failure – theoretically, one might argue that this must be related to the sedimentology of the unit, but the reasons may be elusive (§8.4.1); (2) collecting and sampling failure – the hundred-and-one reasons why specimens were not recognized or collected.

Traditionally, sedimentary successions have been divided into zones (below). For the most accurate biostratigraphic correlation, only changes attributable to evolution and extinction can be used on an international (chronostratigraphic) scale. Although extinction can be reasonably inferred when taxa cannot be found above a certain bed (or level in a well), distinguishing evolutionary from ecological change or migration, when sampling through

Table 8.4 Important groups of biostratigraphic fossils

International	Regional	International	Regional
Precambrian		*Permian*	
stromatolites		ammonoids	miospores
acritarchs		foraminifera	vertebrates[3]
			brachiopods
Cambrian		*Triassic*	
trilobites[1]	trilobites	ammonoids	vertebrates[3]
	archaeocyathids		spores
	acritarchs		vertebrate tracks[3]
	Cruziana		
Ordovician		*Jurassic*	
graptolites	trilobites	ammonites	ostracodes
conodonts	brachiopods		foraminifera
chitinozoans	acritarchs		dinoflagellates
	Cruziana		
Silurian		*Cretaceous*	
graptolites	brachiopods	ammonites	calpionellids
conodonts	acritarchs	coccoliths	ostracodes
chitinozoans		planktonic	belemnites
		foraminifera	radiolarians
		dinoflagellates	Bivalvia[4]
Devonian		belemnites	echinoids
graptolites[2]	brachiopods	inoceramid bivalves	spores and pollen
ammonoids	corals		benthonic
conodonts	cricoconarids		foraminifera
ostracodes	trilobites		
	fish[3]	*Caenozoic*	
	miospores	coccoliths	diatoms
		planktonic	benthonic
		foraminifera	foraminifera
Dinantian		pollen/spores	ostracodes
ammonoids	trilobites	dinoflagellates	bivalves
conodonts	corals	radiolaria	gastropods
miospores	brachiopods		mammal teeth[3]
foraminifera			echinoids
			charophytes
Silesian			
ammonoids	miospores		
conodonts	macroplants		
foraminifera	non-marine Bivalvia		
	ostracodes		

Note:
[1] especially agnostids
[2] in Lower Devonian
[3] in continental facies
[4] e.g. *Inoceramus*, rudists in Tethys

a succession, demands very careful analysis of the evidence. For much of the Phanerozoic, and for many regions, tables of zones are available, and these can be referred to in order to date the rocks. But, consider that such schemes have been produced by careful analysis and judgement of data from many sequences. One is not usually in a position to check the data or concur with the judgement. Neither is it generally possible to be in a position to check that the order of each event in each sequence is the same.

It is an axiom of biostratigraphy that where a succession of zones is the same at widely separated sections, then these are isochronous. If this were not so, then we should not be able to construct our palaeogeographies with much confidence. But it is always wise to consider the possibility of diachronism i.e. that the bioevent did not occur everywhere at the same time. Also be ready to consider that dissimilarity of succession may be due to incomplete collection, or to misidentification.

8.3.1   Procedure in biostratigraphy

Several operations are needed in order to reach a point from which correlation of the section (or core) with other sections (or cores) and with established scales, can begin.

1. Collection of fossils and/or sampling of sediment for microfossils, with reference to the lithostratigraphy (§3.5).
2. Preparation and identification (Appendix B).
3. Keying the lithostratigraphy to the fossil data.
4. If the area is unlikely to be closely linked with well-established stratigraphical scales, then a local biostratigraphy must be generated. The object is to collect material that is going to be stratigraphically useful. Only rarely will more than a fraction of the biota be of such use at zonal level. Attempt to locate such material, and check that specimens are sufficiently complete (e.g. trilobite cephala), adequately preserved, and free of compactional and tectonic distortion for full identification. But a fragment of a compressed Mesozoic ammonite *may* be exceedingly more useful than any amount of most other fossils. Although in practice operations 2–4 are carried out in the laboratory, they are based on evidence collected in the field, and should be done with the participation of the field geologist.
5. Correlation with the regional zones and stages is generally the next operation (Fig. 8.5). For this it is necessary to have a good appreciation of what is meant by the zones and stages, and which groups of fossils are recognized as being most useful for correlation. At most periods of the Phanerozoic, more than one fossil group can be used, e.g. in the Upper Devonian, separate successions of zones can be recognized for ammonoids, trilobites and conodonts (*see also* Fig. 8.2). The boundaries of the zones for each group are seldom coincident. Tables of species ranges are useful. To compare the distributions of fossils in the area under investigation with the regional zonal scheme, the local distributions have to be tabulated. It will be surprising if the local ranges tie-in exactly with the regional zonation. It is often not possible to make more than a subjective judgement. Obviously discard from consideration long-ranging taxa, and see how species with narrower ranges fit.

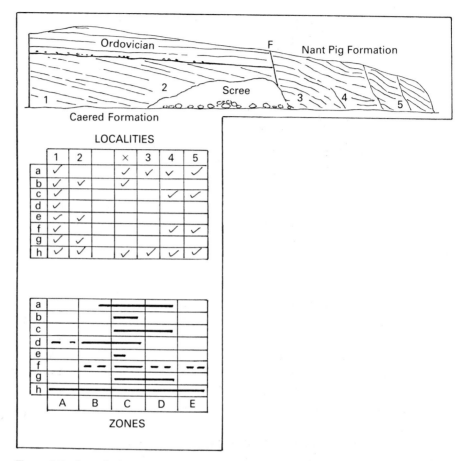

Figure 8.5 A much simplified example of biostratigraphic correlation based on a middle Cambrian sequence in North Wales. Eight fossil species, designated by (a–h) were collected at coastal sites (1–5), and at x, an inland locality of uncertain stratigraphic position, but which is probably older then locality (3). A fault separates the Caered and Nant Pig Formations at the coast. The range of each species (a–h) elsewhere in Britain is shown below with the recognized zones, labelled (A–E). It would seem that the Caered Formation and the lower part of the Nant Pig Formation can be correlated with the lower part of zone (C), the main part of the Nant Pig Formation correlating with the upper part of zone (C) and probably ranging into zone (D). Highest point of cliff 100 m

It is useful to appreciate what the published schemes of zones mean, and there are several ways in which a zone, designated often by a single taxon, can be defined (Fig. 8.6). Zones are the most accurately known division of the biostratigraphic column, and therefore, by implication, of stratigraphic time. But zones, since they concern organisms, cannot have complete global distribution. Consider the distribution of living species today. Even the most widely distributed spores and pollen grains do not get everywhere! But a group of zones (a stage) can generally be used for correlation on an international scale.

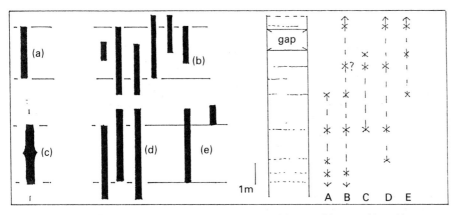

Figure 8.6 (Left) Some types of biozone: (a) total range biozone; (b) assemblage biozone; (c) acme biozone; (d) concurrent range biozone; (e) interval biozone. (Right) A section from which fossils (A–E) are recorded at the points indicated. What types of biozone can be drawn? e.g. fossil A: total range biozone

1. An assemblage biozone is a certain assemblage of taxa which is characteristic of a body of strata, but individual taxa are not confined to the particular interval of rock. This type of biozone is probably the most common type, and it may be designated by one taxon often referred to as the 'index' species. By bad luck this particular taxon may be absent in the section under consideration.
2. A total range biozone comprises the strata associated with a particular taxon.
3. An acme biozone is characterized by an exceptional abundance of a taxon. You may well be suspicious. One thinks of possible ecological controls, but if the evidence shows an abundance, then it should not be ignored on theoretical grounds.
4. An interval zone denotes the strata between two distinctive biostratigraphical horizons, for example, the appearance of taxon *A* and the, stratigraphically higher, appearance of taxon *B*.

Other types of biozones include concurrent range biozone, comprising strata characterized by overlapping ranges of specified taxa. The Lower Palaeozoic graptolite zones, as presently understood, are mainly of this type. Note that combinations of overlapping and total ranges can potentially offer greater stratigraphic resolution.

In using many types of microfossils, lineages are applied. There is often a profusion of material so that ratios of species or abundance peaks can be used. Occasionally, subzones will be encountered. Sometimes these refer to local subdivisions recognizable by the same criteria as zones, but caution should be used in extending subzones widely and assuming isochroniety. Horizons dominated by a single taxon should be correlated with care. In the Chalk (Upper Cretaceous) many such *floods* are of basin-wide extent.

8.3.2   Biostratigraphic boundaries

In describing zones, it has been almost implicit that it is possible to place a boundary between zones. In the field it may be difficult enough to decide where to place a formation boundary. Biostratigraphic boundaries are much more arbitrary. As a well is drilled and cuttings examined, the appearance of a new taxon actually marks its extinction (if the stratigraphy is normal). Disappearances, as drilling proceeds, are generally impossible to define because of contamination by younger fossils from the walls above. Hence, (1) in a well being drilled, position the upper boundary immediately above the appearance of a taxon, or (2) at an outcrop, or completed core, position the base immediately below the first occurrence. These are obviously arbitrary solutions and much depends on the sampling interval, and the time spent on collecting.

8.3.3   An application

Engineers for the Channel Tunnel (Fig. 8.7) have had to design a route that follows closely the Chalk Marl (lower part of Lower Chalk, Cretaceous), which has the best tunnelling characteristics (low permeability, high strength and ease of penetration). The underlying Gault Clay has to be avoided because it is an expanding clay. The design also had to predict the effect of faults, many of which trend normal to the route, throwing the stratigraphy. The Chalk Marl is cyclic (0.5–1.0 m) but in the confines of a tunnel it is not possible to recognize the exact level on the opposite side of a fault. Recourse has to be made to the biostratigraphy and following sponge horizons. Changes in the planktonic and benthonic foraminifera, including abundances, and with local lithological information allow the stratigraphy to be determined to 1–2 metres. Benthonic foraminifera can only be used locally whereas the planktonic forms allow long-range correlation (*see also* §3.5.1). Precision to a few centimetres can be obtained by coring down to the Gault Clay.

8.4   Problems

There are five categories of biostratigraphic problem:

8.4.1   Preservational

If only the fossils were common and evenly distributed through a sequence! If a formation or member appears to be devoid of body fossils, note that many trace fossils, e.g. *Cruziana* (Fig. 6.21) and vertebrate footprints, can be useful stratigraphically. In sandy sediments, shells may have been dissolved out at an early stage. Search for local better preservation or preservations of shells in concretions (*see also* §4.5).

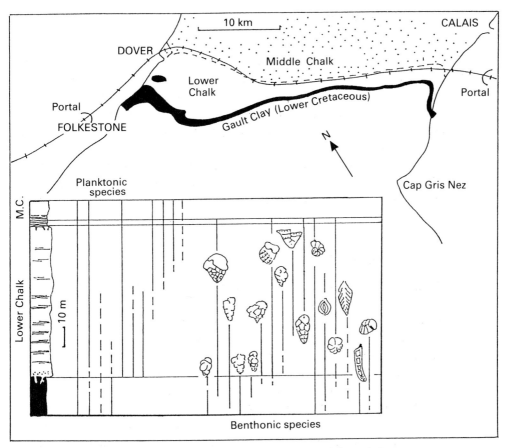

Figure 8.7 Route of the tunnel under the English Channel (England–France). Summary stratigraphic log and indication of ranges of selected planktonic and benthonic foraminifera determined at outcrop (after Destombes & Shepherd-Thorn, 1971; Carter & Hart, 1977). Note extinction event in marls at top of Lower Chalk

8.4.2   Sedimentological (Fig. 8.8)

(a) Dilution. In thick 'expanded' sequences of deltaic sediments or turbidites, the stratigrapher may be asked to subdivide the sequence which may encompass say, one zone. If evolution will not oblige, then recourse must be made to pragmatic alternatives, with perhaps local application only (see Table 8.1).

(b) Condensation and stratigraphic thinning. In contrast, there are sequences which are relatively thin for the amount of time represented. The important point to appreciate in the field is that the sequence may be difficult to locate just because it is thin. Once an outcrop of such beds has been recognized, its nature can be determined. Stratigraphic thinning can arise in a number of ways: the area was starved of sediment (e.g. oceanic rise), or sediment did not accumulate (though

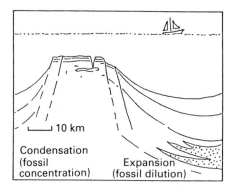

Figure 8.8 Typical situations of fossil condensation, and fossil dilution

sediment may have been temporarily deposited only to be reworked (bypassing)). In the former, sampling may have to be at very close intervals. Condensation can occur by penecontemporaneous erosion or winnowing, leading to an accumulation of heavy shells and concretions. There may be significant stratigraphic mixing or even inversion of the residual fossils. Note carefully differences in preservation mode (Fig. 4.34) and which stratigraphically important fossils also occur in the underlying strata, so as to spot fossils that were contemporary with the condensation event. Problems come when, for instance, three-dimensional phosphatized ammonites lying above an omission surface in the Cretaceous Gault Clay of southeast England, have to be compared with compressed ammonites in the clay below.

(c) It is important in basin analysis to try to evaluate the extent to which sequences are condensed or expanded relative to each other, or whether it is only parts of a sequence that are relatively condensed, indicating fluctuation in rate of subsidence. Pelagic condensed sequences can be readily identified by their closely spaced zones, and thick sequences of turbidites are, virtually by definition, expanded. The problem is often regarded as simply one of accommodation, but criteria that may be used to argue for condensation or expansion, where the biostratigraphy is insufficient for resolution, are, for siliciclastic nearshore and deltaic facies:

Condensed	*Expanded*
Cycles, motifs often incomplete or poorly developed and may only be inferred from facies and intraclasts.	Cycles, motifs complete and distinct. (Tidal, fluvial, deltaic cycles etc.)

Facies

Condensed	*Expanded*
Sharp and generally erosional contacts between facies, ± bioturbated	Gradational change between facies, necessitating arbitrary boundaries
Event beds poorly represented or amalgamated	Event (storm) beds with complete sequence of sedimentary structures, but amalgamated in high energy environments.
Thin bedding typical, cross-bedding poorly developed and represented by lags and toe sets. Channel facies over-represented	Bedding poor to good. Cross bedding often well preserved. Channel facies seldom dominant
Omission surfaces common	Omission surfaces uncommon
Mud layers under-represented, but mud associated with intraclasts and bioturbation structures	Mud layers typical
Coals and lignites poorly represented, but root beds prominent	± thick coals
Pedogenic structures common but thin	Pedogenic structures often associated with coals

Shelly biota

Condensed (time-averaged)	*Expanded*
High corrasion, fragmentation disarticulation, dissociation	Low corrasion, fragmentation disarticulation, dissociation
Shells uncrushed, filled	Shells crushed, ± unfilled, ± dissolved
Minor hydraulic concentrations	Minor hydraulic concentrations
Major transgressive shell beds	Transgressive shell beds generally absent
Hard substrate shells generally over-represented	Infaunal shells common
Shelly fauna over-represented	Shelly fauna diluted
Nekton and plankton under-represented because of taphonomic considerations	Nekton and plankton diluted
Reworked, concretionary fossils common (e.g. phosphatized shells)	Few reworked shells

Autochthonous shells rare and only deeper burrowers represented	Autochthonous shells relatively common
Higher diversity of opportunists and specialists	Opportunists associated with event beds. Specialists under-represented

Bioturbation and ichnofabric

Condensed	*Expanded*
Bioturbation grade high	Bioturbation grade variable, but often confined to tops of event beds, or to particular intervals of cycles
± complex and often indistinct ichnofabric	Ichnotaxa generally distinctive and readily recognizable
Tiering complex because of overprinting	Tiering generally simple especially in event beds
Soles often obscured because of multiplicity of traces, leading to poor bedding definition	Sole traces distinct
'*Ophiomorpha*' rarely pellet-lined	In sandy facies *Ophiomorpha* typically with pellet-lined wall
Escape structures uncommon	Escape structures common
Adjustment structures common	Adjustment structures uncommon

(d) Clastic (Neptunian) dykes represent the filling of cracks that opened on the seabed. One or more events may have occurred, and the crack may not have reopened symmetrically. Neptunian dykes can be large or very thin and may pass into bedding parallel 'sills'. Younger material piped down (stratigraphic leakage) can be difficult to detect. Map out such structures carefully.

(e) Leakage and mixing can also occur when microfossils are washed into burrows, or reworked by bioturbation either downwards or upwards. In both, stratigraphic definition will be reduced.

(f) It is surprising how readily small resistant fossils may be reworked into younger or older sediments e.g. Coal Measure (Carboniferous) spores into Tertiary sediments in southern England, or Pleistocene foraminifera into Eocene clays in Timor. With macrofossils watch for specimens with a mode of preservation that is decidedly different from the associated biota e.g. phosphatic internal moulds (Fig. 8.9).

(g) Intraformational boulder beds and slumps may require careful inspection and sampling. The boulders can be younger or older than the matrix (§2.5).

Figure 8.9 Phosphatized infill of an ammonoid reworked from late Jurassic clays into lower Cretaceous gravels, and extensively bored (Faringdon Sponge Gravels, Oxfordshire). Scale bar = 1 cm

8.4.3   Ecological

There are three types of ecological problem. First, facies fossils are generally recognizable as such by their association with a particular lithology. Some may be long-ranging species that follow the lateral and diachronous facies shift, or crop up each time a facies reasserts itself (as with many bivalve molluscs), but others may exhibit a degree of morphological change, or a different species may appear. Such changes may be useful for local correlation, or even for correlation into another basin (for instance with many brachiopods). Second, even the most widely distributed fossils are to some extent restricted in their distribution, e.g. Jurassic ammonites are virtually unknown in reefal facies. Third, a local facies change may coincide with a disappearance, but this may not be an extinction.

8.4.4   Bibliographical

Stratigraphers and palaeontologists, perhaps more so than other geologists, have to use and, indeed, rely on older literature. The expert will be able to make a shrewd, modern assessment of older descriptions and what an older name is likely to signify now. The stratigraphy was often done in considerable detail, though not often with good facies analysis, and often assuming greater lateral extent of quite thin beds than would be considered reliable today.

8.4.5   Resolution

There is a limit to the fineness of the divisions into which any evolutionary lineage can be separated. Whereas it is true that for some parts of the stratigraphic column, certain groups, e.g. graptolites, ammonites, evolved so quickly that zones of about 0.5 Ma can be recognized, palynological zones of the Jurassic have an average duration of about 4 million years.

8.5   The graphical method of correlation (Fig. 8.10)

The graphical (Shaw, 1964) method of correlating sections is simple to use even in the field, and indeed, should follow construction of cross-sections. A few sheets of graph paper are needed. The method involves plotting graphically the information obtained from two or more sections. The linear relationship, 'regression line', will indicate the best correlation and relative sedimentation rates between the two sections. Any lithological 'event' recorded (tuff, shell bed, storm bed, etc.), as well as biological data, can be plotted. For biological data, plot appearances, disappearances and frequencies. In example (a), the majority of appearances and disappearances cluster about the line $x = y$, but it is readily seen that the ranges of taxa are not coincident. This can be due to four possibilities: (1) incomplete sampling, (2) incomplete exposure,

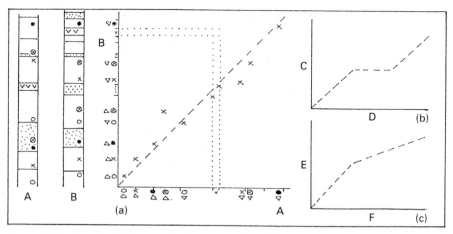

Figure 8.10 Graphical (Shaw) method of correlation. (a) Data from section A plotted against those from section B. Δ – appearances, ∇ – disappearances, of taxa, ● ; o; ⊗; ×; (b,c) see text

(3) actual differences between the two sections due to ecological factors, or (4) actual differences due to geographical factors. The distribution of lithologies is similar in each section, though the tuff horizons are not equivalent (as might be considered on an initial reading). While in the field, it would be sensible to search further for species 'X' in section B, to determine whether higher occurrences may be present.

In example (b), the offset in the correlation line means a sedimentation break, or a fault cuts out part of the sequence in C. The evidence must be in the field! In example (c), there is a marked decrease in sedimentation rate indicated in E relative to F for the upper part of the section. What lithological evidence would be expected?

Note that if two sections correlate well, then sedimentation rates in both are uniform, but not necessarily constant, since one is not correlating actual time. Do a few exercises to help understand the method, by increasing sedimentation rate in one section, and introducing a thrust fault, or slump (with or without erosion of the sole). This method dispenses with zones.

Another 'non-zonal' method of correlation is to compare statistically (e.g. cluster analysis) the types and frequencies of fossils from each sampling point.

When relaxing one evening, ponder over the questions: why can we not dispense with zones? And, might a zonal scheme be still acceptable if it contains unassigned gaps, representing strata from which no zonal fossils have been obtained?

References

CARTER D J, HART M P 1977 Aspects of mid-Cretaceous stratigraphic micropalaeontology. *Bulletin of the British Museum (Natural History – Geology)* **29**: 1–135

CAVELIER C, POMEROL C 1986 Stratigraphy of the Paleogene. *Bulletin de la Société géologique de France* **8**: 255–65

DESTOMBES J P, SHEPHERD-THORN E R 1971 Geological results of the Channel Tunnel site investigations 1964–65. *Institute of Geological Science Report* **71/11** 12 pp

SHAW A B 1964 *Time in Stratigraphy* McGraw Hill, 365 pp (Referred to in many texts on principles of stratigraphy)

Guides to stratigraphic procedure. The two given below are those most widely used.

HEDBERG H D 1976 *International Stratigraphic Guide: A Guide to Stratigraphic Classification, Terminology and Procedure* Wiley, 200 pp

HOLLAND C H *et al.* 1978 *A Guide to Stratigraphic Procedure* Geological Society of London, Special Report No. **10** 18 pp

9

Fossils for structural geologists and geophysicists

'If you choose to ignore life, then stick to the Archaean'

There are six areas where structural geologists and geophysicists can make use of fossils in the field. In many cases the palaeontological evidence provides an independent alternative to structural and geophysical evidence. But the palaeontological evidence may be no more exact than the geophysical evidence, because of lack of documentation (the species has yet to be described), or because of the inherent lack of precision of palaeoecological data (many organisms are ecologically tolerant). Also, relatively few species have a short time range.

9.1   Applications of biostratigraphy (dating by fossils)

Evidence from fossils is needed to determine stratigraphic succession, time gap at unconformities and disconformities, rates of sedimentation (§3.5), and as a check on isotopic ages. In a structurally complex region, dates are needed particularly for structurally isolated units. The approach here is rather different from that required by the stratigrapher for fine correlation. In general, the need is to determine, by any means possible, the age of particular packets of rock. Often only a broad age to stage or series level (§8.1) is required, rather than to the more precise biozone.

Macrofossils may be useful, but distortion and/or recrystallization may render determination difficult or inconclusive. Carbonization of organic matter or replacement by a micaceous mineral is common. The latter is reflective. When examining a suspect mudrock, rotate it to try to spot such preservation. Fossils are generally disposed parallel to bedding, so that it pays to locate bedding and to try to break the rock along it. Some weathering of the rock is usually required for success. Small fossils such as ostracodes, since they accommodate to deformation without significant distortion, can be particularly useful, and may be found on edges of cleavage plates. For instance, finger-print ostracodes are invaluable in the late Devonian and Carboniferous thrust and nappe belt of southwest England. Use a hand lens on reduction spots: there may be a conodont at the centre. Calcareous sediments, even if somewhat impure, or calcareous lenses between lava pillows will be worthwhile collecting for later acid digestion for conodonts (Ordovician–Triassic).

Dating the rocks either side of a major fault at a number of points along the fault will indicate the time, or times, and direction of movement. Anticipate more than one movement, and in different directions.

Fossils in boulders and olistoliths will help date their emplacement time, but note that fossils, like any other sedimentary particle, can be reworked; microfossils so with surprisingly little damage (§8.4.2). Sample each type of clast and also the matrix, since the clasts may be either younger or older than the matrix (§2.5).

An appreciation of rate of uplift of geologically young shallow marine sediments can be obtained when they are found at considerable elevation (e.g. in Appenines), or at depth (submerged reefs), if they can be dated biostratigraphically. The timing of nappe movement may be determined by the progressive younging of the overthrust sediment, caught between faults or in synclinal structures. This is likely to be associated with a progressive younging of the post-tectonic sediment, as with the onset of continental sedimentation from north to south in the Norwegian Caledonides. Whilst the general configuration of an accretionary prism that is still under construction today may be evident from seismic survey, fine biostratigraphy can be used to support or reject a structural hypothesis for an accretionary prism in older rocks. The general younging direction will be towards the site of the former trench and subduction zone, but with each fault-bounded section showing younging in the opposite direction (Fig. 9.1).

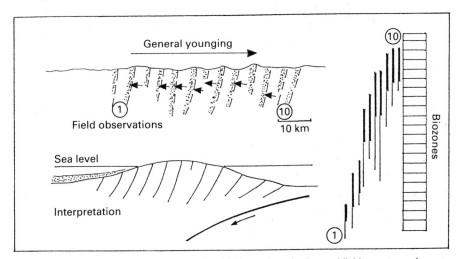

Figure 9.1 Cartoon to illustrate application of biostratigraphy in establishing nature of an accretionary prism with, right, time-stratigraphy of fault slices 1–10. Thicker part of line = turbidites, thin part of line = basinal black shales, chert (after Leggett *et al.* 1979)

9.2   Application to facies analysis

Having determined the stratigraphic succession, the next stage is to find out about the sedimentary environments. The lithology may give sufficient definition, but the biota can be useful, e.g. to decide whether particular turbidites were deposited above or below carbonate compensation depth.

Where structurally separated blocks differ in facies but not in age, it may be possible to estimate original separation distances from palaeoecological analysis. This applies particularly to thrust terranes where shore facies might be brought against deep marine facies. Deformation may be too great to make detailed facies analysis, but try to use broad facies such as shelf, outer-shelf, pelagic facies as a start (Chapter 7). Fossiliferous horizons often have considerable lateral extension making them useful markers, but evaluate the context to make sure that only a single rock unit is involved.

9.3   Biogeographical applications

The recognition of tectonostratigraphic (suspect or displaced) terranes, blocks of crustal rocks accreted onto larger plates, follows naturally from plate tectonic theory. Here is an area where palaeomagnetic and palaeobiogeographic data can be combined. In most examples, it is juxtaposition of fossils which must have been separated originally by many degrees of latitude that provides the clue to the magnitude of the displacement. The examples from the Pacific margins are now classic. For instance, New Zealand comprises at least four tectonostratigraphic terranes (Fig. 9.2) that had accreted into Gondwanaland by the mid-Cretaceous, prior to the area splitting off and moving northwards at the end of the Cretaceous. Unit 3 contains late-Palaeozoic and early Mesozoic faunas, mainly brachiopods, indicating close links with eastern Australia. But unit 1 contains Permian fusulines (spindle-shaped macroscopic foraminifera) generally associated with lower latitudes and suggesting a 5–10° southward movement by transform faulting of an independent block in post-Permian times (McKinnon, 1983). Fossils associated with shallow marine carbonates are most useful because such sediments are most typical of low latitudes. Longitudinal displacement is more difficult to prove. Similarities between tectonically separated biotas can also, of course, be used as evidence for only minor original separation. Much depends on understanding the methods of faunal and floral migration (§6.2.5).

9.4   Geopetal applications

Using body and trace fossils as 'way-up', geopetal, criteria is well known. Since, unfortunately, many strongly deformed sediments are pelagic where grading may be none too obvious and sole marks rare, then any reliable criteria will be useful. 'Way-up' criteria are also required in slumped but relatively

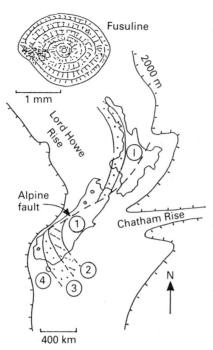

Figure 9.2 New Zealand and adjacent platform. Tectonostratigraphic terraines: (1) Torlesse; (2) Caples; (3) Hokuni; (4) Tuhua and fusulinid in axial section (adapted from McKinnon, 1983). Scale bar = 1 cm

undeformed sediments. The only biological criteria that are wholly reliable are U-burrows (Fig. 7.18) (those normal to bedding), various biogenic sole structures (which never occur on bed tops), predepositional sole traces, and also tiered bed sequences (parallel to bioturbated; §7.3.1). Many trace fossils are much alike on soles and tops, and it would be generally unwise to spend time trying to unravel the situation at deformed sites. The asymmetrical fill of *Zoophycos* (Fig. 9.3) should be used with care, because a reversed asymmetry can occur near to the axial portion of the burrow system (Fig. 6.23). When present, *Zoophycos* is usually common.

Using shell convexity needs care unless there is a clear geopetal indication below the umbrella of the fossil. The attitude of shells reflects the nature of the hydraulic processes (§7.2). Unfortunately, shells are often so flattened as to make them unreliable as 'way-up' criteria. Corals and other organisms with a conical growth form can be used if they are positively in position of life.

9.5   Organic matter and palaeotemperature

All organic material (non-skeletal) undergoes thermal alteration at increased temperature over the range 50–400 °C. The colour attained reflects the temperature and duration at which the temperature is maintained. Colour change is irreversible, but weathering may lighten the objects. The material is slowly carbonized with loss of hydrogen and oxygen. The end result is carbon – black

Figure 9.3 *Zoophycos* in longitudinal section. Note asymmetry of laminae (spreite). White Rock, Wairdridge, Amuri Limestone, Eocene, New Zealand. Scale bar = 1 cm

and highly reflective. But different organisms, because of their differing compositions, exhibit a somewhat varied sequence of colour change associated with increase in reflectivity. Most attention has been given to spores, pollen and plant fragments, in assessing hydrocarbon thermal maturation, but acritarchs, dinoflagellates, chitinozoans, graptolites, insects and fragments of crustaceans and cheilicerates can also be used. Obviously only the larger material will be spotted in the field, and it will be necessary to collect for spores (also useful for biostratigraphy). Organic-walled microfossils can be found in most sediments where the organic material has not been completely oxidized. Collect from fresh darker sediments. Fossil charcoal (fusain) is known from the Devonian and is unchanged with temperature rise.

Conodonts, vertebrate teeth and bones incorporate organic matter within the phosphate, and also exhibit a range of colour change with increasing temperature. White conodonts in shales can be readily spotted, with a hand lens, indicating temperatures in excess of 350 °C.

9.6   Fossils and strain

Figure 9.4 Deformed trilobite, *Angelina sedgwicki* Lower Ordovician, N Wales. Scale bar = 1 cm

Deformed fossils, because they deform like the matrix, can often be used as a means of accurately determining strain in deformed rocks, either when the section is devoid of structures such as pebbles, boudins and folds, or in an independent test. Search for trilobites or brachiopods which still retain symmetry, but have been stretched or squashed, without shear. The trend of the principal axis of the strain ellipsoid can be measured directly. Use can be made of a group of deformed fossils with various orientations, especially when they had original bilateral symmetry (Fig. 9.4). In the field the Wellman (1962) method is useful to determine the strain ellipsoid. Linear fossils such as belemnites, crinoid stems are often boudinaged. Search for specimens that are still straight (without zig-zag), and are thus parallel to the principal strain axis. Measure the stretched length (L_1), and sum the lengths of each original segment, which generally equals the original length (L_0). The elongation is thus $(L_1–L_0)/L_0$. Sketch and photograph specimens showing zig-zag.

Deformed vertical burrows (generally referred to *Skolithos* or *Diplocraterion*, Fig. 7.18, undeformed) have been frequently used in strain analysis. Make measurements on bedding surfaces of the long and short axes of each burrow (from which axial ratio can be calculated), and the angle to reference direction, and plunge of burrow axes. Make sketches and take photographs of marked surfaces normal and parallel to bedding. Include a scale. Note that burrows may have originally varied somewhat from being orthogonal to bedding. When collecting any specimen for later structural analysis, make sure that it is clearly marked with dip, strike and way-up, and that the necessary measurements are recorded: dip, strike (or dip direction), cleavage and cleavage/bedding intersection, lineation. Small fossils, crinoid ossicles and pyrite framboids often display pressure shadows normal to the principal strain.

References

LEGGETT J K, MCKERROW W S, EALES M H 1979 The Southern Uplands of Scotland: a Lower Palaeozoic accretionary prism. *Journal of the Geological Society, London* **136**: 755–70

MCKINNON T C 1983 Origin of the Torlesse terrane and coeval rocks, South Island, New Zealand. *Bulletin of the Geological Society of America* **94**: 967–85

WELLMAN H W 1962 A graphical method for analysing fossil distortion caused by tectonic deformation. *Geological Magazine* **99**: 348–52 plate 16 (Referred to in many texts on structural geology)

Texts

FORTEY R A, COCKS L R M 1986 Fossils and tectonics. *Journal of the Geological Society, London* **143**: 149–50 (Introduction to thematic set of papers)

MCCLAY K R 1987 *The Mapping of Geological Structures* Geological Society Handbook, The Open University Press, Milton Keynes/Halstead Press New York, Toronto 161 pp

PARK R G 1983 *Fundamentals of Structural Geology* Blackie, 135 pp (With useful section of stained fossils)

RAMSAY J G, HUBER M I 1983 *The Techniques of Modern Structural Geology: Vol. 1 Strain Analysis* Academic Press, 307 pp

Appendix A Field equipment and its uses, collecting and preparation

General

Base accommodation
Camping equipment and condition checked
Repair outfits
Water and food sources and supplies
Water treatment
Cooking equipment and fuel

Clothing and footwear (suitable for climate and terrain)
Personal insurance

Location and access

Topographical and geological maps
Aerial photographs
Local access (tide times and vegetation cover)

Indemnity letters required?
Permission to enter land, quarries

Has permission been obtained to collect and (for many countries) will it be possible to keep the material collected, and take it back? Several countries now have stringent rules relating to fossils, and to rocks in general. In Britain hammering and collecting is not allowed at many Sites of Special Scientific Interest (SSSI). In North America and Australia pay close attention to regulations governing the taking of *any* geological material in National Parks, land administered by the Bureau of Land Management and other protected areas.

Equipment

Watch (battery OK?)
Whistle
Torch

Hand lens (attached to string)
Hammers (1 kg, 2 lb) for general use, with chisel end: toffee (4 oz,

First Aid kit
Water bottle
Knife
Survival blanket
High calorie food reserve
Safety spectacles
Sunglasses
Pickaxe, trenching tool,
railroad pick
Measuring tape (2m, 3m or 30 m
length, or meter stick)
Sieves (e.g. 10 mm above 1 mm)
(+ 0.6 mm at top when using sea or
dirty water)
Grain size, sorting comparator
Quadrat (or pegs, string)
Notebook(s), pencils, coloured
pens, water-resistant pens, wax
pencil, type-writer correcting fluid,
adhesive tape (waterproof) carpet
binding tape
Logging sheets

100 gm), plus larger weights
Cold chisels (with handle or glove)
(half-inch, 1 cm) for general use,
brightly painted
Bolster chisel, pinvices
Long chisel (12 inch, 30 cm) for
shales
Pointing trowel
Pincers (to reduce slabs of shale)
Paint brush (for dusting surfaces)

Bucket or bowl
Dilute hydrochloric acid (10%) in
dropper bottle (plastic)
Acidified stain (Alizarin red S)

Packaging: newspaper, tissue paper
(not cotton wool) plastic tubes, bags
(various sizes),

Camera(s) + accessory equipment
Film

Some tips for the field:

1. Sieving is useful for concentrating teeth, small bones from loose sandy
   sediment. Do a trial run and examine what has passed through in a
   bucket or bowl. Do not overload the sieve or attempt to sieve mud or
   clay in the field.

2. Take a photograph of a specimen before extraction, especially if it is
   likely to break up. Photograph trace fossils that are not extractable. With
   fossils of low relief it may be necessary to use a particular time of day.
   Support loose material at an angle similar to the inclination of the sun.
   Wetting (water, syrup) brings out bedding and lamination in terrigenous
   sediments. Always use a proper scale, never coins, hammer, lens cap etc.
   A blob of re-usable adhesive (e.g. Blu-tac), and a partly opened paper
   clip will fix the scale to any surface. Never think that the photograph
   will dispense with the need for written description and sketches.

3. Start each day's notes on a separate page. Indicate date and objectives,
   and the general plan for the day. Organize the notes, even if this means
   jotting down some preliminary notes on a scrap of paper, or on centre

pages (to be removed later). Many notes are of a 'detective' type or ideas to be sorted and analysed later.

4. Use sketches liberally including the support of the photographic record, and recording juxtaposition of blocks, or a broken specimen before removal. Record the location of each sketch and include a scale. Use squared paper as an aid or mark the face into squares. For small (A3–A5) areas place two rulers at right angles on the rock to aid in positioning features. Use simple, not feathery, lines.

5. There is a lot of folklore about collecting. Try to appreciate where the fossils are likely to be found. Also search tip heaps, scree slopes, the end of the conveyor belt, or along the shore where waves have concentrated pyritic fossils, or between boulders, or where rain and wash has concentrated small bones or teeth along gullies. Also examine walls, ploughed fields and immediately under the soil. Be prepared to get down on your hands and knees. Expect the rock and fossil content to change at the boundaries of a quarry or section, including the floor.

6. Blows directed onto hard, massive limestone are rather ineffective. Search around for blocks that are suitable. For heavy blows use the full squared end of the hammer, directed towards yourself and wear goggles (especially for siliceous rock chert or flint). Making outward, glancing blows generally leads to flakes of 'shrapnel' flying off the hammer. When using a chisel to work out the sediment around a fossil, direct the chisel away from the specimen. Never hit one hammer with another. When splitting rocks use a substantial 'anvil'. Tap along lamination with the square head of the hammer. To remove a specimen from a large slab, first chisel out a 'moat' around the specimen. Use tools appropriate to the job.

7. When collecting fossil plants pay particular attention to the mode of preservation, and try to obtain material that shows the full organ and, if possible, organ relationships. Shales and mudstones with plant fossils are often deeply weathered and it is necessary to dig back to fresh material. When collecting from decalcified sandstones only split material that is dry. Examine the edges for an indication of frequency of mouldic preservation.

8. Label specimens before wrapping and add similar information to the wrapping paper or bag and enter data in the field notebook. Mark specimens with way-up and, if necessary, strike and dip. Ring around small and less obvious specimens with a wax pencil. Use permanent marker or scratch temporary numbers. A blob of correcting fluid makes a good base but makes specimens unsuitable for museum curation. Never put part and counterpart into direct contact, and keep thin slabs upright during transport.

9. For field preparation of material the general rule to follow is to carry out the minimum of mechanical preparation. With blocks of fossiliferous muddy sediment, collected for later washing, try to prevent drying out by wrapping in damp newspaper and using plastic bags (preferably double). Spray timber with a fungicide solution. Friable specimens may need supporting on a piece of wood or plastic. Spray fossil leaves with a soluble varnish (e.g. Tricolac) to reduce fragmentation. As a general rule, let covered specimens dry first. Most of the mud will then flake off. With specimens that have broken during extraction, sketch and photograph the distribution of the pieces, mark each and cross-tick adjacent edges, then pack for laboratory reassembling. If an attempt to mend is made use a soluble glue. Plasticine is useful for taking pulls of small specimens but silicone rubber is better, especially if there are parts of the mould that curve away into the rock e.g. spines of a brachiopod. Latex, in an ammonia solution (take care) is useful for casting larger surfaces of low relief. Greater penetration can be obtained if the specimen is moistened with a water/detergent solution and the first few coats should be of solution diluted with a detergent solution. Allow each coat to thoroughly dry. Cellulose acetate is better for making peels of loose sediment. Note that the fumes of acetone are obnoxious and poisonous.

Plaster of Paris (heavy) can be used for support but do not apply it directly to a fossil, unless the surface is covered with waxed paper or aluminium foil. Reinforce with bandages. For displaying bioturbation in white chalk, spray with a solution of methylene blue, or gently smear light oil on a prepared smooth surface. Material that has been exposed to salt spray will have to be well soaked in freshwater. If some preliminary cleaning is attempted, for instance, removing caked and dried mud, take care not to lose important evidence such as the muddy fills to burrows, or the mud below a hardground crevice. For methods of impregnation and consolidation refer to Rixon (1976) and Dowman (1970).

Texts on techniques

BOUMA A 1979 *Methods for the Study of Sedimentary Structures* Kreiger, 458 pp

BRUNTON C H C, BESTERMAN T P, COOPER J A 1985 *Guidelines for the Curation of Palaeontological Materials* Geological Society Miscellaneous Paper **17**

DOWMAN G A 1970 *Conservation in Field Archaeology* Methuen, 170 pp

HOLME N A, MCINTYRE A D (eds) 1984 *Methods for the Study of Marine Benthos* Blackwell, Oxford, 387 pp

KUMMEL B, RAUP D (eds) 1965 *Handbook of Paleontological Techniques* Freeman, 852 pp (Mainly laboratory techniques)

RIXON A E 1976 *Fossil Animal Remains: Their Preparation and Conservation* Athlone Press, 304 pp

Appendix B   Identification

The first question to ask when making comparisons with illustrations, or specimens in museums etc.: Is the mode of preservation the same? If not, then allowance has to be made. This is a more complex topic than it may seem to be, and there are many instances, which strictly are taxonomically unacceptable, of two names having been given, and sometimes still applied, to the same taxon in different preservational states. In fossil plants, for example, *Arthropitys* for anatomical mode and *Calamites* for compressions and pith casts, or *Knorria* for decorticated impressions of *Lepidodendron.* (But note that where different parts of a fossil plant cannot be confidently associated, it is normal to name each part differently – parataxonomy.) In making an identification it pays to have an appreciation of the material. Make sketches (e.g. plan, side views) at a suitable magnification, labelling features that seem likely to be diagnostic. If the identification book has a key or diagnoses peruse these to ascertain what features are diagnostic. Do not be surprised to find diagnostic characters missing from the material to hand because of poor preservation. For example, besides general shape and size a brachiopod diagnosis will refer to internal features such as dentition and muscle areas. Without these, identification often only to genus or family level is possible. Most handbooks of fossils give only the name, age and, possibly, other names that have been applied to the fossil in the past (synonyms) in addition to the illustration. Also handbooks usually carry only a selection of the more common fossils. Local guides, map explanations and accompanying memoirs often have faunal and floral lists for specified localities. Scanning these is helpful.

When an illustration and description that agrees with the material to hand has been identified, ask the question: how close is the agreement? Relatively few descriptions will be accompanied with illustrations covering the complete range of variation shown by a taxon, or even a full (statistical) description of the variation. Check the account to see what variation is referred to. For taxa that are known only by a few specimens, then it is likely that new material, especially if acquired from a new locality, will show some differences. Material of common taxa from well-known localities, is likely to fall within the known range of variation. If the material differs from the published accounts, then it can represent:

1. an extension in the variation of the taxon;
2. a subspecies, characteristic of a geographical area or geological horizon;
3. a taxon that bears only general comparison with a described species, and may be referred to as *Agenus* cf. *beta*.
4. a close new taxon, which may be temporarily designated as *Agenus* n. sp. aff. *beta*;
5. a greater degree of uncertainty may be indicated as *Agenus beta*? or, where specific identification is impossible but the genus appears correct, as *Agenus* sp.;
6. a new taxon *Agenus* sp nov., and the responsibility to ensure that the discovery is made known!

A further problem in identification concerns material from immature or gerontic individuals. For fossils that display continuous growth (brachiopods and bivalves), this is generally readily recognizable, but with fossils of groups that grow by moulting and/or addition of new parts, or by modification of parts, careful consideration is required.

Trace fossils can often be identified with less certainty than body fossils because of incompleteness, and because of the considerable gradation between many ichnotaxa which, in general, is greater than is present between body fossil taxa.

Sources for identification   If there is an indication of the age of the material being identified, this will generally help in the literature search, but can be misleading. And, of course, so often it is the stratigraphic age that is actually required. Each step will generally involve increased time.

1. Start with local guides and local museums.
2. Local and national illustrated handbooks (e.g. as listed below).
   Refer to range charts as a guide, but do not expect them to be infallible.
3. Use the bibliographies in 2 as an introduction to more specific monographs and papers, many of which may be available only from reference libraries in larger cities, universities, etc.

Illustrations and descriptions are published in a wide range of journals and, occasionally, in books, and generally deal with either a specific area or a specific group of fossils, e.g. *Journal of Paleontology, Palaeontology, Paläontologisches Zeitschrift, Alcheringa, Monographs of the Palaeontographical Society*. Some of these journals have cumulative indices which will assist in locating papers. There are other major compilations such as the ongoing *Treatise on Invertebrate Paleontology* (Geological Society of America), where each volume deals with a phylum or class. The *Treatise* is concerned almost entirely with taxa at genus level.

These steps in identification do not mention getting an identification from a local expert, or sending or taking material to experts in research institutions and national museums etc. If you have taken some trouble yourself, have localized the material, and gone some way to understanding the material then, nearly always, the 'experts' are more than willing to help!

Most references to the identification of fossils give little or no ecological information. Palaeoecological interpretation has to be tackled in a systematic way (Chapter 6) working from the individual fossil or species (autecology) to the assemblages and communities (synecology). For fossils with living representatives (even at family level or higher rank) it always pays to assess the extensive literature, videos etc. on the living biota.

Some commonly misidentified groups (see also Appendix D)

ammonoids, gastropods, serpulids	Occasionally, septa in Carboniferous vermetid-like gastropods. Gastropod shell is generally thicker. Coiling in serpulids more or less open ± flanges, and growth often slightly irregular.
scaphopods, serpulid	Tusk-like serpulids in Caenozoic can be distinguished from scaphopods by calcitic shell and closed apex.
brachiopod	Bilateral symmetry, inequivalve (rare exceptions), calcitic shell (can be punctate, pseudopunctate), spines, if present, delicate.
Bivalve	Equivalve, asymmetrical (except oysters, inequivalve; scallops, almost bilateral symmetry, and single (posterior) muscle off-centre).
barnacle plate	Typical diamond-shape, bevelled edge.
massive bryozoans	Orderly thecae.
sponges	Irregular mesh.
bone	Fine mesh, phosphatic.
echinoderm	Very fine, calcitic mesh of stereom.
stromatolites, oncolites	Irregular lamination, no vertical elements.
stromatoporoids	Fine brickwork-like skeleton, ± mamelons.
solenopores (red algae)	Still finer, and more regular, 'conifer-wood in cross-section', cellular structure.
rudists, corals	Small rudists may look like solitary corals, but lack septa.
roots, rhizoliths	Downward bifurcation, with decreasing width. Roots often associated with carbonaceous film and slickensiding.
burrows	Downward bifurcation with uniform width; burrows with menisci or spreite, faecal pellets, variable direction.

modern bone	Greasy and gives off pungent odour when a lighted match applied.
fossil bone	Much heavier, cells ± infilled.
fern leaf	Symmetrical midrib
cockroach wing	Curved midrib
fern leaf seedfern leaf }	Not possible to distinguish without attached seed, or presence of sporangia. Both conditions rare.
angiosperm leaf gymnosperms with angiosperm-like leaf	Angiosperm leaf has blind-ending veinlets. Veins form complete network.
charcoal (fusinite) 'bog oak', jet, vitrinite	Black streak on paper. No streak.

General texts for identification

BRITISH PALAEOZOIC FOSSILS, British Museum (Natural History), 208 pp

BRITISH MESOZOIC FOSSILS, British Museum (Natural History), 208 pp

BRITISH CAENOZOIC FOSSILS, British Museum (Natural History), 132 pp

CHALONER W G, COLLINSON M E 1975 An illustrated key to the commoner British Upper Carboniferous plant compression fossils. *Proceedings Geologists' Association* **86**: 1–44

COLLINSON M E 1983 *Fossil Plants of the London Clay.* Palaeontological Association, London, 121 pp

GILLESPIE W H, LATIMER I S, CLENDENING J A 1966 *Plant Fossils of West Virginia.* West Virginia Geological and Economic Survey, 131 pp

HILL D, PLAYFORD G, WOOD J T 1964–1973 Illustrations of fossil faunas and floras of Queensland. *Queensland Palaeontographical Society*

LINDHOLM R 1987 *A Practical Approach to Sedimentology.* George Allen & Unwin, 276 pp (Also contains key to trace fossils)

MURRAY J W (ed) 1985 *Atlas of Invertebrate Macrofossils* Longman, The Palaeontological Association, 241 pp (With several keys but without archaeocyathids or trace fossils)

SMITH A B 1986 *Fossils of the Chalk* Palaeontological Association, London, 306 pp

TREATISE ON INVERTEBRATE PALEONTOLOGY – ongoing series of volumes covering most groups of invertebrates, including trace fossils and problematica, published by the Geological Society of America and University of Kansas. Besides systematic information many volumes contain detailed descriptions of the biology, physiology and ecology of the modern representatives. (For volumes *see* Chapter 4)

Appendix C   Field statistics

At many fossiliferous sites it is likely that a number of problems will arise requiring statistical analysis. The types of question are:

Is there a preferred orientation to ....?
Is the dispersion of .... clustered?
Do the encrusting organisms show a substrate preference?
Are these two species really distinct?
How similar are these two samples?

A pocket calculator will be required. More sophisticated analysis can be carried out in the laboratory, with the aid of software such as PALSTAT (Ryan & Harper, 1987).

1.   Dispersion (Fig. C1)

Sessile plants and animals may or may not be influenced by the presence of other members of the same species when they become distributed early in life. If they are, then the resulting distribution of individuals will be either 'regular' or 'clumped' (Fig. 6.17), depending on whether they avoid or seek each other out. No influence of any individual on any other will result in a random distribution. Measurements have to be made in the field. The nearest neighbour technique is the one to use.

1. Measure the distance (r) (to nearest mm–cm) between the centre of each individual (or colony) and its nearest neighbour. Lightly mark each sample (but not its neighbour) as it is measured to avoid remeasurement.
2. Calculate the mean (M) of the observed distances

$$M = \frac{\Sigma r}{n \text{ (number of measurements)}}$$

3. Calculate the density (D) of individuals over the sampled area (in mm² cm⁻²)

$$D = \frac{n}{\text{area}}$$

4. Calculate the expected mean distance (E) between neighbours

$$E = \frac{1}{2\sqrt{D}}$$

5. The ratio $R = \frac{M}{E}$ is a measure of the degree to which the observed distances depart from random. A value of $R = 1$ indicates randomness, $R = 0$ indicates maximum aggregation (all fossils at one point), $R = 2.15$ indicates an even distribution.

6. To test the significance of the departure of the mean observed distance from the expected mean distance calculate the standard variate

$$c = \frac{M - E}{0.26136 / \sqrt{nD}}$$

For 95% probability ($p = 0.05$) $c = 1.96$
For 99% probability ($p = 0.01$) $c = 2.58$
Note: Measurements must be taken on individual bedding surfaces only. If the populations are large, take randomly selected individuals, but be sure to measure the distance from each selected sample to the actual nearest neighbour in the whole population.

In the example on Figure C1 the bryozoans are clumped. The *Skolithos*

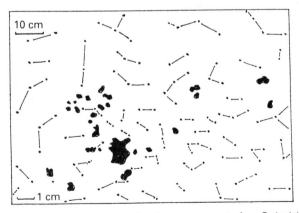

Figure C1 Distribution of encrusting lamellate bryozoa on part of an Ordovician hardground analysed by Palmer & Palmer (1977). Black blobs (10 cm scale). (Other fauna omitted for simplification.) Dots (1 cm scale). Map of part of a quadrat with *Skolithos* burrows (Lower Cambrian, Labrador), with nearest neighbours connected by lines (after Pemberton & Frey, 1984)

burrows show an R value of 1.65, indicating a significant departure from the random towards a regular spacing, probably associated with suspension feeding.

2.   Variation

The eye is particularly good at distinguishing variation and may be said to tend to overestimate diversity. Are two species, which each show considerable variation, really distinct? Although callipers are better, a transparent ruler can be used (inefficiently) to measure parameters, for instance, length (l), width (w) and height (h) of a rhynchonellid brachiopod. Each parameter can be plotted as a histogram. Unimodal plots will suggest that only one species is present. A ratio (w/h) can be plotted against another parameter (l) on a graph. This may distinguish two species as separate clusters of points. Alternatively, plot points for each species (the eye has recognized) and plot the regression lines (Reduced Major Axis) for each: $Y = a + bX$

$$\text{where } b = \frac{\Sigma XY}{\Sigma X^2} \qquad a = \bar{Y} - b\bar{X}$$

where $\bar{Y}$ = mean of Y values

$\bar{X}$ = mean of X values

Are the regression lines significantly different?

3.   Orientation analysis (Fig. C2)

There are three main patterns of preferred orientation in a plane: unimodal, a trend (bimodal) and polymodal.

Measure the trend of the most accurately measured parameter e.g. length of crinoid stems, width of a brachiopod (on a 1–180° scale), and direction e.g. belemnite apex, length of a brachiopod away from umbo (on a 1–360° scale). A circular histogram may be plotted (selecting appropriate class intervals, 10°, 15°, 20° or 30°, depending on the amount of data) by sector area, sector radius or as a spoke proportional to the class frequency. Note that frequencies drawn by sector radii and 'spokes' result in distortion. The radius of each sector area is given by $r^2 = (2Af)/(Nw)$ where r is sector radius, A total area of histogram, f class frequency, N sample size and w class width in radians (1 rad = 180/π degrees).

Alternatively, plot the direction as a circular plot, or trend as a circular plot having doubled the angles (so that North will then be 0° 180°, and South 90° 270°). The circular median direction can then be readily found by searching for the direction that aims for that section of the circumference with the greatest point density, and which divides the sample into two equal parts.

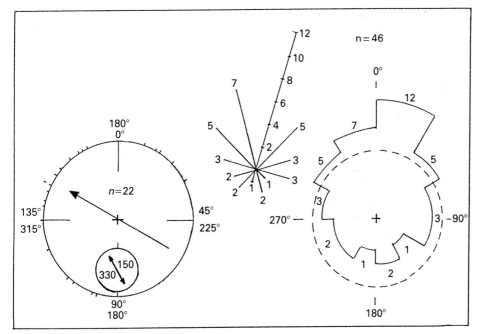

Figure C2 (left) Circular plot of trends of a linear object (e.g. plant stem) with median direction shown, which bisects sample. Here, angles have been doubled so that median trend is 150° (*n*=22). (centre) Spoke histogram of same data as shown right. Data spaced at 30° classes and plotted with radius proportional to class frequency. (right) Circular histogram for 46 measurements grouped in 30° classes and plotted with sector area proportional to class frequency. (Suggestion: to simplify plotting make total area of histogram (A) = 1, so that for a class frequency (f) of 5, then r^2 = (2 × 1 × 5)/(46 × 0.523))

Confidence tests can be carried out later (or see Cheeney, 1983, for Rayleigh test). Where the number of observations is low, and the pattern unclear, use the chi-square test (below) to see whether the data are clumped, compared with the expected even distribution.

On rock faces elongate, cylindrical objects such as crinoid stems or tall-spired gastropods are often viewed end on. There may appear to be a clear preferred trend, but note that orientations above 50° (to the face) give ellipses deceptively close to circular (*see* Fig. C4).

4. Expected versus observed frequencies: chi-squared test

Example questions (see also Lamboy & Lesnikowska, 1988):

1. Are the observed frequencies of valve sorting between samples from 2 beds significantly different? One bed (A) is a bioturbated shelly muddy sandstone, and the other (B) a clearly allochthonous shelly well-sorted fine-grained sandstone.

Observed frequencies:

	Pedicle valves	Brachial valves	Total
Bed A	198	308	506
Bed B	196	258	454
	394	566	960

Expected frequencies:

		Pedicle valves	Brachial valves	Total
Bed A	$\dfrac{394}{960} \times 506 = 208$		$298^{(1)}$	506
Bed B	$\dfrac{394}{960} \times 506 = 186$		268	454
		394	566	960

(1) by subtraction
$506 - 208 = 298$

$$chi^2 = \sum \frac{(\text{observed} - \text{expected})^2}{\text{expected}} \quad i.e.$$

$$\frac{(198-208)^2}{208} + \frac{(308-298)^2}{298} + \frac{(196-186)^2}{186} + \frac{(258-268)^2}{268} = 1.6$$

which is not significant at $p = 0.01$ (χ^2 with one degree of freedom is 6.63) or at $p = 0.1$ ($\chi^2 = 2.71$).

2. Is the amount of breakage between two samples significant? The expected frequencies for the null hypothesis that there is no difference, can be estimated from the combined proportion of broken to whole shells in each sample.
3. Is the proportion of reclining to pedunculate shells in the samples significantly different?

Substrate preference can also be assessed by using the chi-square test to ascertain the probability that one substrate was preferred to another e.g. by encrusters.

1. Make a general inspection of the substrates and encrusters to identify the types of each present.
2. Taking each substrate (of similar area) in turn, record the number (O) of individuals of each different encrusting species.

3. Calculate for each encrusting species, the expected number (E) of encrusters on each substrate

4.

$$E = \frac{\text{total number of substrates of each type}}{\text{total number of each encrusting species}}$$

$$chi^2 = \sum \frac{(\text{observed} - \text{expected})^2}{\text{expected}}$$

5.   A test for small samples (less than 40)

We are often faced with the problem of trying to discriminate between small samples. For instance, might two populations of less than 40 individuals belong to the same or a different taxon? Measurements of the separation between the tubes of U-burrows on two bedding surfaces were made for 20 pairs on each surface. These were set out on a cumulative histogram (Fig. C3). The null hypothesis is that the two samples are of the same taxon and that differences are due to factors such as growth and the variation within each population. The Kolmogorov-Smirnov test statistic (D) for a 95% probability that the two samples are of the same ichnotaxon is given by

$$1.36\sqrt{(N_1 + N_2)/(N_1 \times N_2)}$$

$$1.36\sqrt{20 + 20/20 \times 20} = 0.43$$

On Fig. C3 the maximum separation (observed value of D) is 0.50, much in excess of the expected value. It is thus reasonable to infer that two ichnotaxa are present. If the observed separation had been less, then the null hypothesis could have been supported.

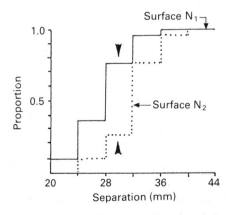

Figure C3   Cumulative histogram of measurements of two populations of U-burrow

N_1, N_2 the separations, measured on the two bedding surfaces
(For $p = 0.01$ (99% probability) D is

$$1.63\sqrt{(N_1 + N_2)/(N_1 \times N_2)}$$

6.   Sampling and rarefaction curves

A sampling curve has been referred to in §3.5. It may sometimes be useful to generate a Rarefaction curve, which might also be called an artificial sampling curve. From a single sample of determined diversity the curve predicts the expected diversity for *smaller* samples. Extending the curve can also give an indication of the way diversity might increase in *larger* samples. Note that Sanders' object, when he introduced the curve, was to demonstrate the marked differences in curve shape that are found for different environments. Example: A sample of 200 individuals contains 20 species, arranged with the abundances:

Table C1.1

Species	n	% sample	cumulative %
1	100	50	50
2	40	20	70
3	24	12	82
4	8	4	86
5	6	3	89
6	4	2	91
7–10	2 each	1% each	95
11–20	1 each	0.5% each	100

Table C1.2

Species	Samples	
	A	B
1	20	50
2	20	35
3	20	10
4	20	4
5	20	1
	$e = 1.0$	0.7

For a sample of 100 individuals one might expect 15 species. Each specimen represents 1% of the sample. From the table above there are 10 species that each represent 1% or more and they account for 95%. One would not expect to find more than 5 more species (100–95).

For a sample of 50 individuals one might expect to find 11 species. Each specimen represents 2% of the sample. From the table there are 6 species that

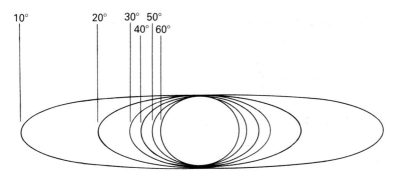

Figure C4 Cross-sections of a cylinder at several angles to the plane of the paper

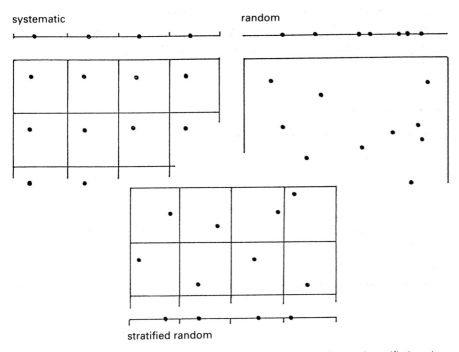

Figure C5 Traverse and areal sampling schemes: systematic, random and stratified random

each represent 2% or more, and these 6 species account for 91%. The residual 9% cannot be expected to yield more than 5 species. The expected number of species is 6 + 5 = 11.

7.   Diversity and species frequency

Is there a significant difference in the way the species are distributed through two samples (collections)? How even are their diversities? While the diversity of two samples may be identical, say with 10 species in each, in one sample each species may be present in equal numbers, while in the other sample two species may account for 80% of the individuals. The two samples may be compared by the Evenness Index:

$$e = -\frac{\sum \frac{ni}{n} \log \frac{ni}{n}}{\log S}$$

where S = number of species

ni = total number of individuals in the ith species

n = total number of individuals

An index that combines richness and evenness (§6.3.2.8) is given by the Shannon Index:

$$(\overline{H}) = -\sum \frac{ni}{n} \log \frac{ni}{n}$$

References

LAMBOY W, LESNIKOWSKA A 1988 Some statistical methods useful in the analysis of plant palecological data. *Palaios* **3**: 86–94

PALMER T J, PALMER C D 1977 Faunal distribution and colonization strategy in a Middle Ordovician hardground community. *Lethaia* **10**: 179–99

PEMBERTON S G, FREY R W 1984 Quantitative methods in ichnology: spatial distribution among populations. *Lethaia* **17**: 33–49

Texts on statistics for palaeontologists

CHEENEY R F 1983 *Statistical Methods in Geology* George Allen & Unwin, 169 pp

RYAN P D, HARPER D A T 1987 *PALSTAT – A Software Package for Palaeontologists and Biologists* Palaeontological Association, Lochee publications

TILL G A R 1974 *Statistical Methods for the Earth Scientist* Macmillan, 154 pp

WRATTEN S D, FRY G L A 1980 *Field and Laboratory Exercises in Ecology* Edward Arnold, 227 pp

Appendix D

Table of skeletal composition, normal depth range, salinity & trophism

Trophism	Depth	Salinity		Mineralogy
			Algae	
Auro-trophs	S³	N	green	A
	—	F²⁵	charophytes	C²⁵
	S²⁶	N–T	red	M (A)
	—	N	coccoliths	C
	—	F–H	diatoms	S
	S	F–H	skeletal stromatolites (blue-green algae)	C, A
		F(H?)	calcipheres (Pal.)	M?
		N	calcispheres (Meso.)	M
			Foraminifera	
	SD	B–T	Textulariina	G¹⁴
21	S	N–H	Miliolina	M
22	SSL	N	Rotaliina	CMA¹⁵
	S	N	Fusulinina	C
21	—	N	Radiolaria	S
Su	S	N	Archaeocyatha	Mg
			Sponges	
	SD	N(B)	Demospongea	S,0³
	S		Lithistids	
Su	SD	N	Hexactinellida	S
	S		Calcarea	CMA
			Sclerospongea	AC²³
			Stromatoporoidea	
			Mesozoic	A
Su?	S	B?–T	Palaeozoic	M²
			Coelenterata	
			Hydrozoa	A
			Scleractinia	A
			Octocoralallia	MA
Mc	S¹⁶	N	Rugosa	C
			Tabulata (27)	C (A)
	SD		(conularids)	P
			Bryozoa	
Su	S(SL)	T	Cheilostomata	M (A)
			rest	C

Trophism	Depth	Salinity		Mineralogy
Su	S(SL)	N	Brachiopoda	C (M)
		N–T	most inarticulates	P
			Mollusca	
W	S(SD)	F–H	Gastropoda	A⁴
Dp	S	N	Scaphopoda	A
Cr¹⁷	S–D(17)	N	Cephalopoda	A⁵
			Bivalvia	24
Dp	S	N	protobranchs	A
			oysters	C⁶
			mytilids	B⁷
			pinnids }	B⁸
			pterids }	
			pectens	C⁹
Su¹⁸	S	19	anomids	C¹⁰
			limids	B¹¹
			hippuritids (rudists)	B
			Chama	A¹²
			rest	A
W	SD	F–H	Annelids	OPG
Su		F–T	serpulids	M (A)
			Arthropoda	
20		N	trilobites	C
W	S	F–H	ostracodes	C M
Su		T	cirripedes	C
W		F–H	decapods	CP¹³
Su		F(B/N)	branchiopods	C P
W			insects, arachnids	O
Cr	S	F–B	merostomes	O
			Echinodermata	
Su			stemmed	
Dp/Bw(Su)	S/SL	N	echinoids	M
Cr			'starfish'	
Su/Cr			brittlestars	
			Vertebrata	
			bone, teeth	P
W	—	—	eggshell	C
			otoliths	A
?	SSL?	N	conodonts	P

KEYS: A – dominant (A) – minor

Trophism

Su	Suspension	– taking food from suspended matter in water } without need to break up particles
Dp	Deposit	– taking food from substrate }
BW	Browser	– scraping plant material from surfaces or chewing vegetative matter (grades into deposit feeding)
Sc	Scavenger	– eats dead organisms (grades into deposit feeding)
Cr	carnivore	– takes live prey
Mc	microcarnivore	– supplemented in hermatypic Scleractinia by commensal autotrophs (zooxanthellae)
W	wide range	

Mineralogy

C	Calcite
M	Mg calcite: >4mol% MgCO₃
A	aragonite
B	bi-mineralic, C, A
S	siliceous
O	organic
P	phosphatic
G	agglutinating

Salinity

N	stenohaline normal marine (30-40ppt)
T	euryhaline greater tolerance (20–50 ppt)
B	brackish
H	hypersaline
F	freshwater

Depth (marine benthos)

S	shelf (to 200m)
SL	slope (to 2000m)
D	deep
SSL	shelf to slope
SD	shelf to deep

Notes:
1. *Solemya* chemisymbiont suppl.
2. Affinities of some in dispute
3. Spongin or S+spongin, S readily replaced by C, typically ⪢50 m.
4. Most extant archaeogastropods (*Pleurotomaria, Patella Trochus*, neritids) with outer C, Inner A
5. Aptychus, guard, rhyncholites C
6. A only in ligament and myostraca
7. Variable: entirely A or outer prismatic C and inner A (nacre)
8. Outer C prisms, inner nacre A
9. C (foliated) (some A in middle layer)
10. Outer C foliated, assd. with A (cross laminar)
11. Inner A (cross laminar)
12. 2 species with C
13. C as reinforcement
14. Habitat derived grains (quartz, sponge spicules, other foraminifera); cement organic, calcareous or FeO
15. Most larger M and globigerinids C
16. Hermatypic Scleractinia typically < 20 m, 20–25°C.
17. Ammonoids may have had wide range of depth and trophism. Implosion depth is useful depth criterion (e.g. *Nautilus* 700–800 m).
18. Mainly Su except Dp *Tellina* and related genera; *Teredo* wood
19. Marine except for F unionoids and M-B-F mytilacids
20. Probably W
21. Pseudopodial trapping
22. Some supplemented by commensal autotrophs (zooxanthellae)
23. Plus spicules S
24. C only in some epifaunal groups (or in groups with epifaunal ancestry). Tube lining of *Teredo* C, others A or?
25. M in brackish environments. Some modern forms tolerate hypersalinity
26. Mostly shallow, but to 100 m and greater, Articulated inter-tidal
27. A in Tetradiidae

Principal Source: Lowenstam H H, Weiner S 1989 *On Biomineralization* Oxford 324 pp.

Appendix E

Generally recognized chronostratigraphic divisions

Europe	(common to both)	N. America

CENOZOIC (TERTIARY)	QUATERNARY
	  Holocene (Recent)
	  Pleistocene
	NEOGENE
	  Pliocene
	  Miocene	

Messinian
Tortonian Helvetian   *Clovellian*
Serravallian   *Ducklakean*
Langhian   *Napoleonvillean*
Burdigalian   *Anahuacan*
Aquitanian   PALEOGENE (Pg)
                 Oligocene

Chattian   *Chickasawhayan*
Rupelian   *Vicksburgian*
                 Eocene

Priabonian   *Jacksonian*
Bartonian   *Claibornian*
Lutetian   *Sabinian*
Ypresian
              Palaeocene

Thanetian
Danian   *Midwayan*

(65)
MESOZOIC
CRETACEOUS (K)

Maastrichtian ⎫   Gulfian
Campanian ⎬ Senonian   *Navarroan*
Santonian   ⎪   *Tayloran*
U   *Coniacian* ⎭   *Austinian*
Turonian   *Woodbinian*
Cenomanian   Comanchean
                                          Washitan
Albian   (Gault)   *Fredericksburgian*
                                          Trinitian
Aptian   Coahuilan
L   *Barremian*   *Nuevoleonian*
Hauterivian
Valanganian ⎫ Neocomian
Ryazinian   ⎬ (Wealden)   *Durangoan*
             (Purbeckian)   *Berriasian*

(144)
JURASSIC (J)

                 (Malm)   U ⎧ *Portlandian* ⎫ *Volgian* ⎫ *Tithonian*
                                ⎨ *Kimmeridgian* ⎬
                                ⎩ *Oxfordian* (Corallian)
                 (Dogger)   M ⎧ *Callovian*
                                ⎨ *Bathonian*
                                ⎪ *Bajocian*
                                ⎩ *Aalenian*
                 (Lias)   L ⎧ *Toarcian*
                                ⎨ *Pliensbachian*
                                ⎪ *Sinemurian*
                                ⎩ *Hettangian*

Appendix E (cont.)

Europe	(common to both)	N. America

(213)
TRIASSIC (Tr)

U { *Rhaetian*

M, L middle columns:
    Norian   } (Keuper)
    Carnian

  M { *Ladinian* } (Muschelkalk)
       Anisian
  L   *Scythian*   (Buntsandstein)

(248)
PALAEOZOIC
PERMIAN (P)

U { *Tatarian*   Thuringian   U { *Ochoan*
   Kazanian   (Zechstein)   *Guadalupian*
   Kungurian

L { *Artinskian*   Saxonian
   Sakmarian   (Rotliegendes)   L { *Leonardian*
   Asselian   Autunian   *Wolfcampian*

(286)   UPPER
SILESIAN (Si)   CARBONIFEROUS (C)   **PENNSYLVANIAN (Pen)**
Stephanian   Kawvian
  Stephanian   *Virgilian*
  Cantabrian   *Missourian*
Westphalian   Oklan
  Westphalian D } Moscovian   *Desmoinesian*
            C }   *Bendian*
            B }   Ardian
            A } Bashkirian   *Morrowan*
Namurian   MISSISSIPPIAN (Mis)
  Yeadonian   Tennesseean
  Marsdenian   *Chesteran*
  Kinderscoutian
  Alportian
  Chokierian
  Arnsbergian
  Pendleian

               LOWER
DINANTIAN (Di)   CARBONIFEROUS (C)   *Chesteran*
Brigantian
Asbian
Holkerian   } Viséan   *Meremecian*
Arundian   Waverlyan
Chadian   (Culm)   *Osagian*
Courceyan   Tournaisian   *Kinderhookian*

(360)
DEVONIAN (D)
                      Chautauquan
                      Conewangoan

U { *Famennian*   (Farlovian)   *Cassadagan*
                      Senecan
        (Old Red Sandstone)   *Chemungian*
  Frasnian   *Fingerlakesian*

Appendix E (cont.)

	Europe	(common to both)	N. America
M {	Givetian		Erian
			Taghanican
			Tioughniogan
	Couvinian	Eifelian	Cazenovian
			Ulsterian
	Emsian	(Breconian)	Onesquethawan
L {	Siegenian	Pragian (Dittonian)	Deerparkian
	Gedinnian	Lochkovian (Downtonian)	Helderbergian

(408)
SILURIAN (S)
Ludlow   Cayugan
   Whitcliffian
   Leintwardinian
   Bringewoodian
   Eltonian
Wenlock   Niagaran
Llandovery   Medinan
   Telychian
   Fronian
   Idwian
   Rhuddanian

(438)
ORDOVICIAN (O)
Ashgill   Cincinnatian
   Hirnantian   *Richmondian*
   Rawtheyan   *Maysvillian*
   Cautleyan   *Edenian*
   Pusgillian
Caradoc   Champlainian
   Onnian   *Mohawkian*
   Actonian   *Trentonian*
   Marshbrookian   *Blackriveran*
   Longvillian
   Soudleyan
   Harnagian
   Costonian
Llandeilo   *Chazyan*
Llanvirn
Arenig   Canadian
   Fennian
   Whitlandian
   Moridunian
Tremadoc

(488)
CAMBRIAN (Є)
Merioneth   Croixian
                                          Trempealeauan
                                          Franconian
                                          Dresbachian

Appendix E (cont.)

Europe	(common to both)	N. America
St. Davids		Albertan
Comley	*Lenian*	
	Atdabanian	Waucoban
	Tommotian	
(540)		
(EDIACARIAN)		
PRECAMBRIAN		
	Vendian	
	Riphean	

References

HAQ B U, EYSINGA T W B van 1987 Geological Time Scale (4th edn) Elsevier, Amsterdam (chart)

HARLAND W B, COX A V, LLEWELLYN P G, PICKTON A G, SMITH A G, WALTERS R 1982 *A Geological Time Scale.* Cambridge University Press, Cambridge, 131 pp

HARLAND W B, ARMSTRONG R L, COX A V, CRAIG L E, SMITH A G, SMITH D G 1990 *A Geologic Time Scale 1989* Cambridge University Press, 263 pp

Appendix F Glossary

(*See also* environments Fig. 7.1, bedding Fig. 3.1 and coal types, Fig. 4.12)

References

ALLABY A, ALLABY M 1990 *The Concise Oxford Dictionary of Earth Sciences* Oxford,
410 pp (6000 entries)

BATES R L, JACKSON J A (eds) 1987 *Glossary of Geology* (3rd edn) American
Geological Institute, Alexandria, Virginia (contains 37 000 terms)

Active fill: Material actively passed into burrow by animal, often pelleted
(= backfill)

Actuopalaeontology: The study of organisms and their taphonomy in modern
environments in order to better understand fossils

Aerobic: Conditions are aerobic in the presence of free oxygen (> $1.0 \, \text{ml} \, l^{-1} \, O_2$)

Allochthonous: A fossil is allochthonous if it has been transported from the
environment in which it originally lived (cf. autochthonous)

Allogenic: Generated externally

Anaerobic: Devoid of oxygen (< $0.1 \, \text{ml} \, l^{-1} \, O_2$)

Anoxic: Devoid of oxygen

Association: A consistently recurring group of fossils that may be assumed to have
lived together, but which is only part of a total palaeocommunity.

Astogeny: Development by asexual reproduction leading to branched colonial
organism such as colonial (compound) corals, bryozoans and graptolites

Autochthonous: A fossil is autochthonous if it is found in the place where it
originally lived (cf. allochthonous)

Aut(o)ecology: The study of the (palaeo)ecology of an individual organism or
species

Autogenic: Self-generated

Autogenic succession: An ecological succession resulting from factors inherent to
the community, self-generated

Basin analysis: Analysis of the extent and context of facies within a basin

Benthonic: Relating to the substrate

Biocoenosis: An association of living organisms

Biogenic sedimentary structure: See trace fossil

Bioimmuration: Encrustation of soft tissues by skeletal organism, to leave a
natural impression of the soft tissue

Biostratinomy: See stratinomy

Bioturbation: Process of sediment mixing by organisms and breakdown of primary
sedimentary structures

Body fossil: The remains or representation of a whole or part of an animal or plant, e.g. shell, impression of a jellyfish or skin, charcoalified plant

Boring: Biogenic sedimentary structure excavated into a hard substrate by chemical means or by grinding

Bouma cycle: A predictable succession of lithologies, hydraulically related, that make up a turbidite bed

Buildup: A body that has original topographic relief. In this text autochthonous buildup (Fig. 3.3) is used in a wide sense for an accumulation of dominantly autochthonous organisms. Though generally evident, a framework may be obscured by taphonomic processes

Burrow: Biogenic sedimentary structure in soft sediment below the sedimentary surface

Coelobite: Cavity-dwelling organism (applied in palaeontology especially to encrusting and nestling organisms within borings)

Commensalism: A relationship between two organisms in which one benefits, but is not injurious to the other

Coquina: Detrital limestone composed chiefly or wholly of mechanically transported skeletal material

Community: A group of animals and/or plants that live together. A palaeocommunity is a community of animals and/or plants that are assumed to have lived together in the past

Crypt: A cavity

Cryptic: Refers to cavity dwelling organisms

Cycle: Sedimentary sequence reflecting recurrent, repeated events, so that conditions at beginning and end are the same

Derived fossil: Skeletal material derived (reworked) from an earlier sedimentary cycle

Diagenesis: The changes (chemical, physical and biological) that occur in a sediment after its initial deposition, and during and after lithification, but excluding weathering and metamorphism

Diastem (non-sequence): A relatively short interruption in sedimentation

Disconformity: Unconformity without structural discordance

Disjunct: Where a taxon exists in separate areas

Dysaerobic: Very little free oxygen (0.1 to 1.0 ml l^{-1} O_2)

Ecology: The study of the interactions of organisms with one another and with the environment

Edaphic: Relating to soil

Endemic: Restricted geographically

Energetics (ecological energetics): Energy transformations within ecosystems (community and environment)

Epifauna: (a) Fauna living upon rather than below surface of the seafloor, (b) Fauna living attached to rocks etc

Epilith: Any organism attached to rock or a bioclast

Epizoan: Any (animal) organism attached to a *living* organism (adj. epizoic) (syn. epizoite)

Euryhaline: Organisms tolerant of a wide range of salinity. Salinity range 30–40‰

Eutrophic: Body of water with high level of plant nutrients

Fabric: Spatial arrangement of particles in a sediment

Facies: Unit of rock defined by its total geological character (geometry, lithology, sedimentary features and fossil content). This is the way it is used here: in an observational and descriptive sense. But note that it is also often used in a genetic sense e.g. turbidite facies or, in an environmental sense e.g. shallow marine facies, or in a tectono-sedimentary sense e.g. post-orogenic facies

Facies sequence analysis: Analysis of the vertical organization of facies

Firmground: Stiff but uncemented substrate

Formation: The formation is the primary local lithostratigraphical unit. It is mappable and should possess internal homogeneity with distinctive lithological features. Thickness should not be a determining feature

Form genus: A taxon of convenience used in the classification of fossils of problematic relationship. The taxon may be unassignable to a higher taxonomic category

Fossil: Any remains, trace or imprint of a plant or animal that has been preserved, by natural processes, in the Earth's crust since some past geologic or prehistoric time

Fossil-Lagerstätte: See Fossil-Ore

Fossil-ore (Fossil-Lagerstätte): Any fossiliferous site yielding an unusual amount of palaeobiological information because of the excellence of preservation (e.g. soft tissue), or because of the abundance of material. An old, disused definition is: fossiliferous iron ore

Guild: A group of species that exploit the same class of environmental resources in a similar way, without regard to taxonomic position, e.g. construction guild of reefs

Hardground: A sedimentary non-sequence during which the substrate became lithified. Typically recognized by encrusting or boring organisms

Heterochrony: Different rates of reaching maturity: neoteny, acceleration. Change in timing of ontogenetic events

Homoeomorphy: Superficial resemblance between taxa, but dissimilarity in detail. (NB: biologists define more narrowly, relating only to species)

Hummocky cross-stratification: Type of cross-stratification associated with laminae that dip at low angles and with little or no preferred orientation. Sole generally planar but succeeding bounding surface hummocky. Associated with storm events

Hypersaline: Denoting salinity substantially greater than that of normal sea water: >40‰

Ichnocoenosis: Trace (fossils) formed by a co-existing benthic community

Ichnofabric: Sediment texture and internal structure due to bioturbation and bioerosion. The ichnofabric integrates the taphonomic and stratinomic aspects of the trace fossils with the primary fabric

Ichnofacies: Recurring associations of trace fossils on all scales

Ichnofossil: See trace fossil

Inertinite: Macerals characterized by high carbon content

Infauna: Aquatic organisms that live essentially below the substrate surface

In situ: In place in the rock at outcrop (or core), as distinct from in scree, soil or clitter

Kerogen: Acid insoluble organic residual

Lebensspur (plural-lebensspuren): See trace fossil

Lumachelle: An accumulation of shells, especially oysters. See coquina

Maceral: Microscopic constituents of coals

Member: A part of a formation recognized by a particular lithological peculiarity. It is not necessarily mappable

Meniscate: Active burrow fill of thin, dish-shaped laminae formed behind advancing animal

Mesohaline: Salinity between 18‰ and 30‰

Meteoric: Water of recent atmospheric origin

Mutualism: A relationship beneficial to both organisms

Necrotic processes: Processes of death

Nektonic: Swimming

Neomorphic: Skeletal replacement by another mineral (or the same mineral) without an intervening void stage

Neoteny: Acceleration of sexual maturity with retention of many juvenile characters. (Syn. paedogenesis)

Obrution: Smothering (rapid burial) (§6.3)

Oligohaline: Salinity between 0–5‰

Oligotrophic: Water body with deficiency of plant nutrients

Ontogeny: Development of an individual organism (*see also* astogeny)

Opportunists: Said of organisms able to rapidly colonize vacant niches

Ordering: Sequence of burrows/borings in a bed recognized by cross-cutting relationships. Burrows relate to a single substrate surface. Absence of ordering indicates contemporaneous colonization

Palaeocommunity: See Community

Palaeoecology: The study of the interactions of organisms with one another and with the environment in the geological past

Paleosol: A fossil soil. NB: Most fossil soils do not show roots but they do provide indication of modification of the material by near surface (terrestrial) organic, chemical and physical processes

Palynofacies (analysis): The quantitative analysis of the environmentally typical residue of a sediment, after elimination of the mineral phase by acid digestion. The residue comprises spores, pollen, foraminiferal tests-linings, microplankton, plant and animal cuticle, inertinite, vitrinite and amorphous organic matter

Parataxonomy: System of classification in which parts of an animal or plant (or work) are classified separately (*see* form genus)

Parautochthonous: A fossil is parautochthonous if only little post-mortal transport has occurred and burial is, usually, in the same environment (biotope)

Pelagic: Swimming or floating organisms

Phenotypic (variation): Variation (in morphology) due to environmental factors, e.g. crowding

Phreatic: Water in saturated zone (= ground water)

Phyletic gradualism: Evolutionary process or pattern in which morphology changes gradually and continuously

Placer: Mechanical concentration

Planktonic: Floating

Polyhaline: Salinity between 18–30‰

Prefossilization: Diagenetic change to fossil material in the primary sedimentary cycle, before final burial

Punctuated equilibrium: Evolutionary process or pattern in which morphologic change is concentrated within a relatively short interval of a species range. A period of stasis follows

Regression: A change that brings nonmarine, nearshore or shallow water conditions over deeper offshore conditions, an extension of land areas

Rhythm: Term used interchangeably with Cycle

Riparian: Living beside rivers or streams

Sapropel: Plant material degraded under anaerobic conditions

Sequence: Conformable succession of facies arranged in a predictable manner. Generally bounded by sharp junction (boundary)

Sequence stratigraphy: The application of seismic stratigraphic interpretation techniques to sedimentary basin analysis

Spreite (plural – spreiten): The successive walls of a displaced burrow, produced as animal shifts broadside through sediment

Stenohaline: Organisms tolerant to a narrow range of salinity

Stratinomy: The processes between death of an organism and its final burial. Also applies to trace fossils e.g. at hardgrounds

Synecology: The study of the relationship between (palaeo)communities and their environments

Taphocoenosis: An assemblage of fossils characterized by similar modes of preservation, and, normally, similar stratinomic history

Taphofacies: Suites of sedimentary rock characterized by particular preservational (taphonomic) features

Taphonomic feedback: Change in community structure due to changes in the nature of the substrate, associated with dead skeletal material

Taphonomy: All the changes that occur to an organism (or its work) between death and discovery as a fossil (Fig. 4.1)

Taxonomy: The theory and practice of the classification of animals and plants

Tempestite: The sedimentary product of a tempest, typically displaying a sequence of sedimentary depositional structures attributable to a storm and its wane. The base is typically erosional and the top gradational and/or bioturbated

Texture: Relative proportion of, especially, different size classes of particles. The physical appearance of a rock

Thanatocoenosis: A burial assemblage, where each fossil is in the position it died (autochthonous)

Tiering: Trace fossils and body fossils are tiered above and below the substrate surface. Tiering is a synonym of the biological term stratification

Time-averaging: Mixing of skeletal components by physical or biological processes over a relatively short time span, thereby averaging out temporary, stochastic (random) fluctuations within a community

Toponomy: Mode of preservation of biogenic sedimentary structures

Trace fossil: Evidence of the activity of an organism in ancient sediments e.g. fossil burrows and footprints. Also includes fossil faecal pellets and coprolites

Transgression: A change that brings deeper water conditions over shallower water conditions, or a spread of sea over land areas

Trophism: Nutrition involving metabolic exchange in the tissues

Tube: Burrow with prominent lining

Turbidite: The product of a turbidity current, typically demonstrating a sequence of sedimentary depositional structures associated with a decelerating current. Degradation is typical at the base. The top is typically gradational with or without bioturbation

Unconformity: A substantial stratigraphical break in the rock record, with or without structural discordance (*see also* diastem)

Vadose: Water of aerobic zone

Appendix G

Appendix G Checklists

A *Outcrop*
1 Location
2 Dip/strike
3 Lithology (colour, grain size/sorting)
4 Bedding (cm – m)
5 Cleavage(s), jointing
6 Strength, hardness

B *Autochthony*
1 Natural orientation
2 Skeletal dissociation (L, R)
3 Shell attitude
4 Size sorting
5 Bedding orientation
6 Breakage/damage
7 Lithology/matrix
8 Clustering
9 Adjacent sediment

C *Ecological factors*
1 Bathymetry
2 Temperature
3 Salinity
4 Substrate
5 Oxygenation
6 Inter-relationships
7 Abundance
8 Diversity
9 Dispersion

   (plants)
1 Temperature – precipitation
2 Light
3 Soil
4 Biological inter-relationships

D *Buildups*
1 Autochthony
2 Ecological role:
    binding, encrusting
    stabilizing
    frame builders
    bafflers
    loose sediment
       producers
3 Growth form
4 Growth stage:
    domination
    diversification
    colonization
    stabilization
4 Growth rate
5 Local relief
6 Matrix
7 Cavities
8 Diagenesis

E *Allochthonous skeletal accumulations*
1 Local (cm – m)
    Bedding stringers, pavements,
       current shadows, clusters
    Winnowed layers
    Gutter fills, burrow fills
    Desiccation struct., solution fissures,
       tectonic fissures
2 Extensive (m – km)
    Lags, relict accumulations,
    Storm beds, turbidites
    Pelagic accumulations

F *Trace fossil taphonomy*
1 Mode of preservation epichnia,
    endichnia, exichnia, hypichnia
    (convex, concave), undertrack,
    ichnoclast
2 Endogenic, exogenic
3 Sediment consistency
4 Ordering
5 Tiering
6 Variation in preservation mode
7 Identification, description

G *Body fossil taphonomy*
1 Modes of preservation: original shell,
    skeleton, replacement, encrustation,
    external impression, internal
    impression (core), composite
    mould, crystal core, tectonic
    distortion, compaction cast
2 Skeletal dissociation
3 Epizoan, epilith
4 Preferred orientation (gravity, light,
    food, current)

H *Approach to basin analysis*
1 Lithology
2 Bedding/stratinomy
3 Facies analysis
4 Facies sequence analysis
5 Basin analysis
(6 Global analysis)

Index

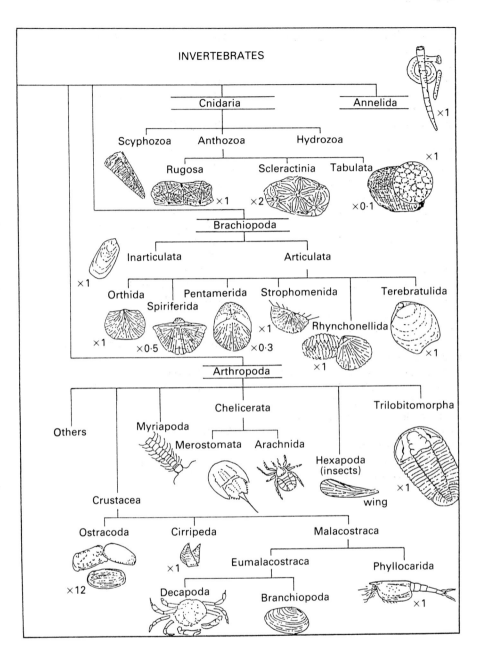

INVERTEBRATES

Cnidaria

Annelida

×1

Scyphozoa   Anthozoa   Hydrozoa

Rugosa   Scleractinia   Tabulata

×1   ×2   ×0·1   ×1

Brachiopoda

Inarticulata   Articulata

×1

Orthida   Pentamerida   Strophomenida   Terebratulida

Spiriferida

×1   ×0·5   ×1   Rhynchonellida   ×1

×0·3   ×1

Arthropoda

Chelicerata   Trilobitomorpha

Myriapoda

Others   Merostomata   Arachnida

Hexapoda
(insects)

×1

wing

Crustacea

Ostracoda   Cirripeda   Malacostraca

Eumalacostraca   Phyllocarida

×1

×12   Decapoda

Branchiopoda   ×1

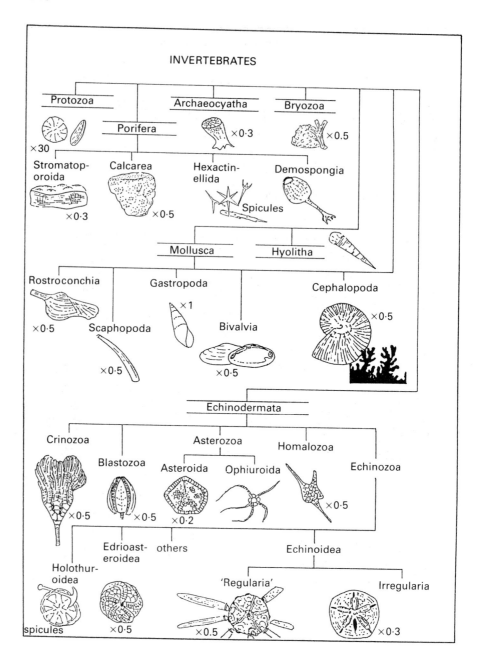

INVERTEBRATES

Protozoa

Porifera

Archaeocyatha ×0.3

Bryozoa ×0.5

×30

Stromatoporoida ×0.3

Calcarea ×0.5

Hexactinellida

Demospongia

Spicules

Mollusca

Hyolitha

Rostroconchia ×0.5

Gastropoda ×1

Scaphopoda ×0.5

Bivalvia ×0.5

Cephalopoda ×0.5

Echinodermata

Crinozoa ×0.5

Blastozoa ×0.5

Asterozoa

Asteroida ×0.2

Ophiuroida

Homalozoa

Echinozoa ×0.5

Holothuroidea spicules

Edrioasteroidea others ×0.5

Echinoidea

'Regularia' ×0.5

Irregularia ×0.3

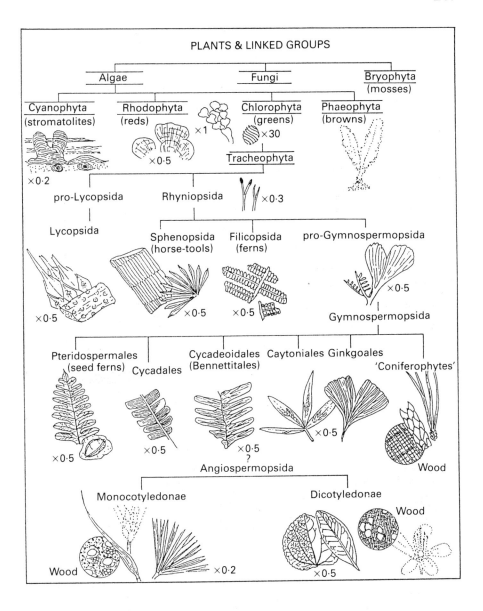

PLANTS & LINKED GROUPS

Algae   Fungi   Bryophyta (mosses)

Cyanophyta (stromatolites)   Rhodophyta (reds)   ×1   Chlorophyta (greens)   ×30   Phaeophyta (browns)

×0·5

×0·2

Tracheophyta

pro-Lycopsida   Rhyniopsida   ×0·3

Lycopsida   Sphenopsida (horse-tools)   Filicopsida (ferns)   pro-Gymnospermopsida

×0·5

×0·5   ×0·5   ×0·5

Gymnospermopsida

Pteridospermales (seed ferns)   Cycadales   Cycadeoidales (Bennettitales)   Caytoniales   Ginkgoales   'Coniferophytes'

×0·5   ×0·5   ×0·5   ×0·5

Wood

?

Angiospermopsida

Monocotyledonae   Dicotyledonae

Wood

Wood   ×0·2   ×0·5

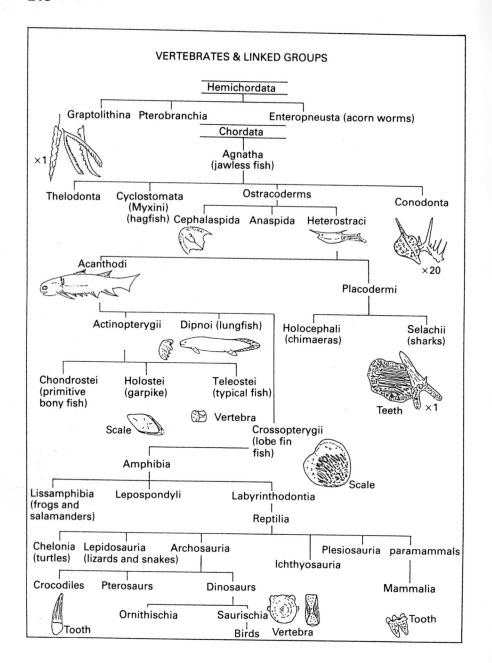

VERTEBRATES & LINKED GROUPS

Hemichordata

Graptolithina Pterobranchia   Enteropneusta (acorn worms)

×1

Chordata

Agnatha
(jawless fish)

Thelodonta   Cyclostomata   Ostracoderms   Conodonta
             (Myxini)
             (hagfish) Cephalaspida Anaspida Heterostraci

Acanthodi   ×20

                                   Placodermi

             Actinopterygii Dipnoi (lungfish)   Holocephali   Selachii
                                                  (chimaeras)   (sharks)

Chondrostei   Holostei   Teleostei
(primitive   (garpike)   (typical fish)
bony fish)

                              Vertebra   Teeth   ×1

Scale

                                           Crossopterygii
                                           (lobe fin
                                           fish)

             Amphibia   Scale

Lissamphibia   Lepospondyli   Labyrinthodontia
(frogs and
salamanders)   Reptilia

Chelonia   Lepidosauria   Archosauria   Plesiosauria paramammals
(turtles) (lizards and snakes)   Ichthyosauria

Crocodiles   Pterosaurs   Dinosaurs   Mammalia

             Ornithischia   Saurischia
                                                                           Tooth
Tooth   Birds   Vertebra